THE PARABLE OF THE BEAST

The Parable of the Beast

John N. Bleibtreu

The Macmillan Company, New York

The Macmillan Company, New York
Collier-Macmillan Canada Ltd., Toronto, Ontario
PRINTED IN THE UNITED STATES OF AMERICA

Acknowledgment is gratefully made to the following copyright holders for permission
to reprint the following: Abridgement of Chapter 22, "A Behaviorial Sink" by John B.
Calhoun, from *Roots of Behavior*, edited by Eugene L. Bliss, MD. Reprinted by permission of Hoeber Medical Division, Harper and Row, Publishers © 1962.

Illustrations reprinted by courtesy of: *Scientific American*, illustration from John Tyler
Bonner "How Slime Moulds Communicate," August 1963 Vol. 209, No. 2. Copyright ©
1963 by Scientific American, Inc. All rights reserved. Photographs reprinted courtesy
of: Dr. Kenneth B. Raper, Dr. John T. Bonner and Dr. G. Gerisch.

"To the rational animal, the same act is according to nature and according to reason."

MARCUS AURELIUS

"Certainly no man hath yet determined what are the powers of the body: I mean that none has yet learned from experience what the body may perform by mere laws of nature, considering it only as a material thing, and what it cannot do without the mind's determination of it. For nobody has known as yet the frame of the body so thoroughly as to explain all its operations."

SPINOZA

Contents

Introduction

The voice of the turtle is heard in the land, heard in all the arts—in literature, painting, and music—and in the voices of men and women speaking to one another. It is not the voice of the dove, that sweet and melancholy sound which the translators of the Authorized Version presumably had in mind; it is the croak of isolation and alienation issuing from within a vault of defensive armor—the voice of the reptilian turtle. This armor we wear—the armor of technology separating us from the rest of the natural world—has created us lately in the condition of exiles. Nature exists within as well as without, and we are become, therefore, exiled from ourselves. The style of the catatonic has become the style of Everyman.

The more we control nature through our multitudes of ingenious contrivances, the unhappier we seem to become without ourselves both individually and collectively. We can make night into day, summer into winter; we can distract ourselves from that most natural of all sufferings, the delicious pangs of solitude—but nature invades us nonetheless.

This book is about the nature within. The things that appear in nature, the birds and bees and other animals, appear also in this book. But they are actors, not subjects. The subject of the book is Man, and all the various animals that fly and crawl and swim through its pages are no more the subjects of these tales than were the fox and geese of Aesop the subjects of his. In Aesop's book the behavior of his animals was, right on the surface, unbelievable. The reader was expected to suspend his understanding of reality and become preternaturally credulous, to become again as he was when a child. The reader of this book is not expected to render up to it his critical judgment. The facts reported here are attested to be "factual" by sincere and competent observers using the most objective procedures their imaginations could construct and the best observation equipment available to them.

Still, the facts they report can only be understood as parables. Some of the parables appearing here point toward conclusions which are widely accepted as capable of being derived from the facts as they are presently known. Some of them are aimed at conclusions which many professional biologists would regard with a cold eye. But I believe the purpose of a parable is to expand the imagination, not to contract it.

In this generation Biology is coming of age as Queen of the sciences. In the last two decades progress has been achieved which makes some of the work of the 1940's appear almost as if it had been accomplished in the last century. Previously Physics and Mathematics reigned jointly over that body of hard knowledge we in the West call science. But Physics and Mathematics treat of the material world, a world of dead things, a world apparently lacking in volition. Today the borderline between this world and the volitional world of living things is growing ever more blurred. Science in the nineteenth century had every reason to be vain, even arrogant, over its conquests of ignorance. Vanity is perhaps the most contagious of diseases of the soul, and laymen who had no great acquaintance with science became, nevertheless, infected by its vanities. They acquired a great and false faith in the ability of science to comprehend the mysteries. Western science has been traditionally unwilling to collect the mysteries within a structured metaphysical or religious system of thought. This is quite right and proper, since metaphysical speculation has an unfortunate readiness to crystallize quite rapidly into dogma. Dogmatic rigidity is the enemy of truth, for it lays claim to truth, while the *reality* of truth lies buried somewhere in the paradox of its nonexistence. But as they are extended into mythologies, metaphysical systems allow mankind the means to abide with mystery. Without a mythology we must deny mystery, and with this denial we can live only at great cost to ourselves.

It seems to me that we are in the process of creating a mythology out of the raw materials of science in much the same way that the Greeks and Jews created their mythologies out of the raw materials of history.   I feel strongly that this is not only a legitimate, but a necessary process.

In order to avoid cluttering the text with footnotes, the author has identified the sources of all quotations, data, and statistics in the Notes *to the book.   Each quote or study mentioned is identified by the appropriate page number in the text where the reference or quote is made.   In addition, this section serves as a brief but thorough bibliography of the scientific papers and monographs, books, articles, and reports which went into the making of the book.*

PART ONE

The Individual

[*1*]

The Moment of Being

The cattle tick is a small, flat-bodied, blood-sucking insect with a curious life history. It emerges from the egg not yet fully developed, lacking a pair of legs, and sex organs. In this state it is still capable of attacking cold-blooded animals such as frogs and lizards, which it does. After shedding its skin several times, it acquires its missing organs, mates, and is then prepared to attack warm-blooded animals.

The eyeless female is directed to the tip of a twig on a bush by her photosensitive skin, and there she stays through darkness and light, through fair weather and foul, waiting for the moment that will fulfill her existence. In the Zoological Institute, at Rostock, prior to World War I ticks were kept on the ends of twigs, waiting for this moment for a period of eighteen years. The metabolism of the insect is sluggish to the point of being suspended entirely. The sperm she received in the act of mating remains bundled into capsules where it, too, waits in suspension until mammalian blood reaches the stomach of the tick, at which time the capsules break, the sperm are released and they fertilize the eggs which have been reposing in the ovary, also waiting in a kind of time suspension.

The signal for which the tick waits is the scent of butyric acid, a substance present in the sweat of all mammals. This is the only experience that will trigger time into existence for the tick.

The tick represents, in the conduct of its life, a kind of apotheosis of subjective time perception. For a period as long as eighteen

3

years nothing happens. The period passes as a single moment; but at any moment within this span of literally senseless existence, when the animal becomes aware of the scent of butyric acid it is thrust into a perception of time, and other signals are suddenly perceived.

The animal then hurls itself in the direction of the scent. The object on which the tick lands at the end of this leap must be warm; a delicate sense of temperature is suddenly mobilized and so informs the insect. If the object is not warm, the tick will drop off and reclimb its perch. If it is warm, the tick burrows its head deeply into the skin and slowly pumps itself full of blood. Experiments made at Rostock with membranes filled with fluids other than blood proved that the tick lacks all sense of taste, and once the membrane is perforated the animal will drink any fluid, provided it is of the right temperature.

The extraordinary preparedness of this creature for that moment of time during which it will re-enact the purpose of its life contrasts strikingly with probability that this moment will ever occur. There are doubtless many bushes on which ticks perch, which are never by-passed by a mammal within range of the tick's leap. As do most animals, the tick lives in an absurdly unfavorable world—at least so it would appear to the compassionate human observer. But this world is merely the environment of the animal. The world it perceives—which experimenters at Rostock called its *umwelt*, its perceptual world— is not at all unfavorable. A period of eighteen years, as measured objectively by the circuit of the earth around the sun, is meaningless to the tick. During this period, it is apparently unaware of temperature changes. Being blind, it does not see the leaves shrivel and fall and then renew themselves on the bush where it is affixed. Unaware of time it is also unaware of space, and the multitudes of forms and colors which appear in space. It waits, suspended in duration for its particular moment of time, a moment distinguished by being filled with a single, unique experience; the scent of butyric acid.

Though we consider ourselves far removed as humans from such a lowly insect form as this, we too are both aware and unaware of elements which comprise our environment. We are more aware than the tick of the passage of time. We are subjectively aware of the aging process; we know that we grow older, that time is shortened by each passing moment. For the tick, however, this moment that precedes its burst of volitional activity, the moment when it scents butyric acid and is thrust into purposeful movement, is close to the end of time for the tick. When it fills itself with blood, it drops from its host, lays its eggs, and dies.

For us humans, death seems to come in a more random fashion. Civilized as we pretend to be, we know very little about time and the transformations of living things that occur within it. We know little about growth and aging, which, it would seem, are stimulated by the passage of time which streams around us carrying us with it; and about death, or the termination of biological time.

Unlike astronomers and physicists who were and are concerned with totally abstract and relative concepts of time, biologists, until very recently, considered time, as nearly all of us do in the course of our daily lives, to be a linear dimension. An hour is as much a linear measure as a mile. Even when biology began considering time abstractly, as inferred from the geological and fossil record, it was treated in linear fashion, as a historical process. Though it is a commonplace observation that history repeats itself, no systematic approach, other than the linear study of history, has ever been found useful. And so the process of evolution is seen as being linear-historical. But today, over one hundred years after the publication of Charles Darwin's *The Origin of Species*, we still maintain a curiously double view of ourselves as organisms developed by historical time. We are well aware that, like all living things on earth, we have been touched by a kind of recurrent wind that has blown on other things at other times. We like to think of ourselves as being unique and individual personalities, and as such, occupying a unique niche in time. Yet common to a great many religious systems is some notion of reincarnation; a poetic statement to the effect that we have, in fact, been touched by this historical, recurrent wind, that we are only in one stage of a general evolutionary process, and like the tick we are possessed of a limited perceptual system. We live within a specialized *umwelt*, not, to be sure, quite as specialized as that of the tick, but equally selective of stimuli. This double, near-schizophrenic view of what we can "know" of ourselves and the world is most clearly displayed when we limit our acknowledgment of our evolutionary heritage to merely the formal, or structural part of ourselves—the physical self. We admit, for example, that our five fingers probably developed from paws. But what of the inner self? Here the double vision becomes insupportable as each view departs from the other. The general structure of contemporary social and political institutions depends on the conviction that, being rational creatures, we can force the future into the mold of our expectations—into eras of peace and plenty—by exercising reason. We manage, by exercising the most credulous and hopeful optimism, to overlook the inexorable, repetitive bestiality of human history. We are animals: we can see it in our conduct as clearly as we can see the anatomical connections to the beast as we lie bare-boned and gutted on the autopsy table. As the horrid history of this century—a history so deformed by atrocity as to make a parody of all humanist assumptions regarding our capacity for rational thought—as this history winds itself into the future it is most suitable and appropriate for us to remind ourselves of our limiting antecedents.

Over the course of our evolutionary journey we have gained an incredible intellectual heritage; we have contrived a system of comprehending causal relations and recreated our ecological environment, but we have lost as well. On the simplest, most pragmatic level this loss becomes obvious when we must provide special "survival schools" for downed airmen deprived of their civilization so they may *learn* how to survive—simply to exist, to sustain life—in a world which their primitive ancestors found sustaining, and which sustained those previous stages of man's metamorphosis for hundreds of millions of years. This scrim of culture has blurred our perception of the natural *umwelt*; this blurring loss of contact is symbolized by a lack of knowledge of how to do certain things, such as make a fire without matches. But the real loss is far more extensive. We have lost whole areas of experience. We are, in effect, numb to them, as the tick, having evolved the specialized conduct of its life, is numb to the very passage of time. We have lost, in general terms, a sense of intimacy with the cosmos, an innate knowledge of our belonging to and with all the living forms that swarm thickly on and through the surface layers of our planet. We have lost our ability to exist harmoniously within the solar system and within the cosmos at large.

This sense of loss has become acute in the West during this century, and the principal burden of contemporary art forms consists of proclaiming this sense of alienation. Previous to our time man has sustained himself by extending the limitations of his sensible world to a mythological cosmology, which at least corresponded to the realities of apprehension. But as Nietzche announced toward the end of the nineteenth century, God is dead. Myths are dead, and with the death of myth the world became random, accidental, meaningless.

It is not only the artist who describes this malaise. Werner Heisenberg, the great German physicist, writes of his conversations with Niels Bohr in Copenhagen during the 1930's, using those terms of anguish and doubt which are familiar to us from the vocabularies of theologians and philosophers. But Heisenberg was speaking of his discipline, physics, which according to the scientific myths of the nineteenth century was considered the "hardest," most doubt-free and positivistic of all the natural sciences. He writes: "I remember discussions with Bohr which went on through many hours till very late at night and ended almost in despair; and when at the end of the discussion I went alone for a walk in the neighboring park I repeated to myself again and again the question: Can nature possibly be as absurd as it seemed to us in these atomic experiments?"

Heisenberg concluded that nature is not absurd, but that, because it is necessary for us to contrive *artificial* perception devices in order to "know" it, we intrude the deficiencies of our perception into those objects and processes we wish to know about. He writes: ". . . the observation plays a decisive role in the event, and the reality varies whether we observe it or not. . . . It is very important to realize that our object [of study] has to be in contact with the other part of the world, namely the experimental arrangement, the measuring rod, etc., before, or at least at the moment of observation. . . . This influence introduces a new element of uncertainty . . . and since the device is connected with the rest of the world, it contains in fact, the uncertainties of the microscopic structure of the whole world."

This book will attempt a brief account of one modern effort to manufacture part of a new, sustaining myth which corresponds to reality. As man evolved the symbolic language of speech, he lost, to a great extent, his ability to converse with his environment in the language of direct experience. This loss has occurred because we have trained our intellect to intervene between ourselves and experience. We have lately discovered there are forces abroad in the world to which animals respond. At one time or another in our evolutionary history we, too, responded to these forces. They now remain mostly vestigial like the twenty-eight day lunar cycle of human female menstruation. We once lived much more intimately with the moon and experienced its influence more directly upon our activities. This new mythology which is being derived from the most painstaking research into other animals, their sensations and behavior, is an attempt to re-establish our losses—to place ourselves anew within an order of things, because faith in an order is a requirement of life.

The tick has "faith" even though it lacks a sense of taste. It has the faith that if it lands on a warm object reeking of sweat and finds it filled with fluid, this fluid will be mammalian blood. In the Zoological Institute, at Rostock, the "faith" of certain ticks was not rewarded, for they were presented with warm bladders filled with neutral water, or destroying acid, and they drank deeply and mistakenly in their faith. But the probability is, in the *natural* order of things, that if a tick came upon a warm object smelling of sweat, this thing would be a mammal, a mammal filled with blood.

Once we then acknowledge the world as being subjectively perceived, we must also acknowledge that the world varies with an individual's given sensibilities. Are there sensibilities of which we are not aware? Do we receive a "knowledge" of the world and the order of things from sources of sensibility that are as yet inaccessible to the intellect? Every myth, including this new one, maintains that we do, and that what we call intuition is a response to these sources of intelligence; that the acknowledgment of the "reality" of intuition is an indispensable part of the dialogue of experience.

The subjective exploitation of intuition, the mystic search, has generally been alien to the secular temperament of the West since the end of the Renaissance. There has been no consensus of validity to the reports of such experiences. If they are judged at all, they are judged by aesthetic standards, not by the pragmatic standards of "reality."

The post-Renaissance standards of objective analysis, however, have been applied to subjective states, but not, strangely enough, by psychologists—by zoologists. It is no accident that one of the great contributions to our understanding of human sexuality was made by a zoologist, Alfred C. Kinsey.

We are probably too close to it to be able to unravel the peculiar labyrinthian course of intellectual history, yet it would seem, off hand, as if the prevailing egalitarian ethos of the United States and Russia were responsible for the direction taken by the study of psychology in those two countries. There was a very strong commitment to the proposition that all men were quite literally *created* equal. As a result the study of psychology centered on the mechanics of the learning process, so that, by proper teaching, a truly egalitarian society could be created. In Russia the major figure soon appeared to be Ivan Pavlov who achieved eminence by his proclamation of the principle of the conditioned reflex. He was able to teach dogs to salivate on the artificial stimulus of a bell rather than the functional stimulus of the odor of food. In the United States John Watson dominated animal behavior studies. A generation of his followers constructed hundreds of models of mazes and puzzles, all roughly based on the principle of reward or punishment. Through these structures untold numbers of caged and captive animals were run, in hopes that a key might be found to an understanding of what learning (and presumably knowledge) is comprised. Intellectually, both systems of psychology were based on mechanical theoretical models. Behavior resulted from input, output, feedback systems as though it were the result of some kind of computer operation which took place in the brain or the nervous system.

Little emphasis was placed on the "given" elements underlying modes and patterns of behavior. Formerly these given elements were all lumped in the general category of "instinct" and during the heyday of Behaviorism, "instinct" was a naughty word. It was considered a totally plastic entity almost infinitely capable of being modified by experience. Few zoologists labored under such delusions. Kinsey, for example, did not for one moment believe that the result of living under a certain set of rules or experiential conditions altered the human sexual "instinct" as it expressed itself in behavior.

A great deal of valuable information was obtained from the Watson and the Pavlov studies, but since they were motivated exclusively by pragmatic considerations, they contained their own inbuilt limitations of application. The frightening social and political consequences of communities based on these models of behavior were lampooned by Aldous Huxley in *Brave New World*, and by George Orwell in *1984*.

In Europe animal behavior studies took an entirely different direction. They were conducted by zoologists, not psychologists. The emphasis of post-Darwinian zoology has been on evolution, on answering the very simple, basic question: How did living things get to be the way they are now? As to how forms acquired their present appearance, part of the answer is hopefully to be found in the fossil record. At least the sequence of the appearance of transitional forms is to be found there. But behavior cannot be fossilized. It can only be deduced; from the shape of an animal's teeth, for example, one can deduce whether it was herbiverous or carnivorous, and from that, deduce, in general, its style of life—whether it was a grazer, browser, or hunter. But the hard answers to these questions come from witnessing the acts of life themselves, not from the circumstantial evidence which is often misleading.

And so in Europe animal behaviorists did not work in laboratories, with animals that for generations had been reared in cages. They went out into the field with their binoculars and notebooks, and when they did bring animals into the laboratory for detailed studies, they were wild animals, and every attempt was made to keep the conditions of their captivity as close to those of the wild as possible. Since these studies were motivated by an abstract curiosity and had no immediately obvious applications, they received little popular attention. At first only people directly concerned with animals—zookeepers, conservationists, game wardens, and so on—read the publications and found any empirical uses for the new knowledge.

The initial approach was quantitative—like Kinsey's with human sexuality. What is the range of overt behaviors, and the range of their frequencies, which exist within a given population of animals? Concurrent with this approach, but also receiving less public attention at the time, was an attempt to enter into the subjective world of the animal being studied.

Over the course of time this subjective approach became a discipline, that is, it acquired its own jargon and a sect name for itself—Ethology. It appeared first in continental Europe, for on the continent a certain resistance to the "objective" rigors of orthodox Darwinian theory still prevailed. There remained in France some vestigial influences from the thought of Jean Baptiste Lamarck, who preceded Darwin and believed that animals exercised volitional control over the direction of their evolution. As George Bernard Shaw (one of Lamarck's very few British supporters) wrote, "the great factor in evolution is use and disuse. If you have no eyes and want to see, and keep trying to see, you will finally get eyes. If, like a mole or a subterranean fish, you have eyes and don't want to see, you will lose your eyes. If you like eating the tender tops of trees enough to make you concentrate all your energies on the stretching of your neck, you will finally get a long neck, like the giraffe. This seems absurd to inconsiderate people at the first blush; but it is within the personal experience of us all that it is just by this process that a child tumbling about the floor becomes a boy walking erect. . . ."

In Germany Lamarckism did not seize as great a hold on the popular imagination as it did in France, where, then as now, the holding of an intellectual view first propounded by a Frenchman was not so much a matter of judgmental choice, as it was an act of personal vitality and an affirmation of national grandeur. Another factor worked in Germany—the intellectual disposition toward romantic idealism, which tended always to weight subjective "irrational" factors underlying behavior more heavily than the objective forces of reason. Again it is no accident that the two greatest proponents of the irrational, unconscious motivations that produce human behavior, Sigmund Freud and Carl Jung, appeared from within this German-speaking world. It seemed, therefore, to many continental biologists, that the Darwinian theory of evolution gave too little consideration to the subjective elements of the animal's inner being.

To some extent this criticism was well taken, for Darwin's

ideas derived originally from Thomas Malthus who was primarily an economist.   Darwin's famous phrase, by which his theory is popularly understood—"the struggle for existence"—he borrowed from Malthus's *Essay on Populations*.   Like Linneaus, who preceded him, Darwin was obsessed with form, with the external appearance of animals.   On reading Darwin, one begins to feel almost as if he envisioned the environment acting upon successive generations of animals as though it were an abrasive—as sand and water shape rocks.   When Darwin looked upon an environment, he did not see what the Germans saw, a perceptual universe—an *umwelt*—he saw a literal patch of geographical space and a span of linear time.

Malthus admits that his famous essay grew out of some discussions that he had with his father "respecting the perfect ability of society." Like Pavlov and Watson he was an optimist, hopeful of developing some theoretical key to improve the human condition by mitigating the abrasion between the environment and the organism.

The man who coined the term *umwelt*, who examined the perceptual system of the cattle tick, and who is considered by many to be the father of ethology, was an eccentric Baltic Baron named Jakob Johann von Uexküll.

A photograph taken in his seventies shows him short and robust, impeccably and formally dressed with a heavy golden chain slung in two portentous curves across his waistcoated stomach—a fine, full Bacchic stomach, a fitting foundation for the barrel chest and bull neck that rise above.   Atop a perfectly round head he wears his white hair cut short in the German pussy-willow style.   His eyebrows bristle, and below them curve a great pair of exuberant moustaches which swing across his cheeks almost joining his sideburns in the Franz Joseph tradition.   He smiles broadly off to one side of the camera, but like Charlie Chaplin, his smile is tainted with tragedy.   It comes as no surprise to discover that he was a friend of Rainer Maria Rilke.   He is a mixture of styles; the dress and pose and barbering are straight out of the nineteenth century, but the ambiguous, ironical smile is the mark of the twentieth. His life and thought conform to this mixture of styles apparent in his appearance. He was born in 1864—at the same time that Bismarck rose to power —to a distinguished Baltic family, which traditionally supplied officers for the armies of the Tsar, the King of Sweden, and the German Imperial General Staff. He attended a school in Reval where the rector was a man who later became the father of Wolfgang Koehler, another student of animal behavior who was one of the founders of Gestalt psychology. Perhaps the power of Rector Koehler to arouse curiosity and philosophical speculation worked on both his own son Wolfgang and on von Uexküll with sufficient strength to bend Jakob Johann away from his historical family career of militarism. At any rate, well before the turn of the century von Uexküll read Immanuel Kant and became fascinated by the twin phenomena of time and space. And time—inasmuch as it represents life itself, the duration of life —is the first concern of any living organism.

Von Uexküll was concerned with the given elements which underlie behavior, with the inherited dispositions that prompted animals to do the things they did. It was all very well that Pavlov succeeded in teaching a dog to salivate at the sound of a bell rather than upon perceiving the odor of food. Of more interest to von Uexküll was the odd fact that the sense of smell is a system by which we and animals perceive the presence of certain special chemical molecules freely floating in air. There are some chemical molecules which directly affect our survival—carbon monoxide, for example—but we cannot perceive the presence of carbon-monoxide molecules even when they abound in lethal concentrations. What Pavlov had done was convert the dog's recognition of important elements in its *umwelt* from a chemical sense receptor to an accoustical one. Von Uexküll was concerned with the evolution of sense receptors and the *umwelts* which proceeded from them—the subjective environments of men and animals—and how men and animals manipulated the conduct of their lives within these *umwelts*.

He considered time to be the most objective entity within the environment of all living things and he concentrated on producing evidence that the passage of time and living creatures' relationships to it were subjectively perceived and genetically determined.   All living things respond to the passage of time.

The prevailing concept of time today implies that time can best be viewed as a function of the Second Law of Thermodynamics.   The First Law of Thermodynamics states that energy can change its form, but not its quantity.   The Second Law states that any given process in this universe is accompanied by a change in magnitude of a *quantity* called entropy.   This entropy is a quantity, not a quality.   It is the *amount* of heat-reversible exchanged from one part of the universe to the other.   Entropy is also the *measure* of the randomness, or the lack of orderliness in the system.   The greater the randomness, the greater the entropy.

For most practical purposes it is enough to imagine that if the temperature be constant, the measure of decay in a system is a measure of its age—or the passage of time since the system originated.

For example, the time required for a given quantity of pure uranium 238 to "decay" into lead 206 is 4,560 years.   Thus time can be measured not only by planetary movements, but also by rates of decay, when these rates are known to proceed in orderly fashion.

Von Uexküll, however, was convinced that "objective" time measurements—those which used as reference points, happenings in nonliving systems, such as planetary movements (entropy-change measurements were not known to him)—could not be transposed in terms of subjective meaning.   He believed that evolutionary studies should include not only comparative anatomy, but also comparative behavior.   Yet it is most difficult when studying behavior to avoid the Walt Disney syndrome of anthropomorphism.   Von Uexküll chose the most objective entity common to all animal environments—time.   Space, though equally as much a constant as time in the environment of every animal, seemed to him far less susceptible of study. Space to an earthworm is obviously an entity completely different from space to an albatross. In addition a very considerable portion of every animal's behavioral repertoire consists of self-protective activities—activities designed to avoid predators and to increase the span of linear time available to the animal and its offspring. The whole Darwinian concept of "adaptation" and fitness involves mechanisms by which the animal acquires linear time to establish itself and its species.

Believing, then, that an understanding of the way in which animals perceived time was crucial to an understanding of the evolution of behavior, von Uexküll began as best he could with the technology available to him. He noted, for example, that trout seem to react more rapidly to stimuli than snails, and accordingly devised an experiment. He trained a trout to associate the color gray with a food reward and to associate the colors black and white with a punishment, an electric shock. He placed the fish in a tank with three blank walls. In one of these walls was a small window. To train the fish, he showed behind this window a gray card and placed food in a tray beneath it. When he showed a black or a white card through this window, the fish was shocked. During the actual experiment itself the black and white cards were replaced with a rotating disk painted in alternate patches of white and black, which showed through the window. As the disk spun ever faster, the black and white patches eventually merged into gray. At this point the trout approached the tray for food. Until the disk reached a certain speed of rotation, the trout had been able to note the alternation of black and white and remained far removed from the food tray. By means of this experiment von Uexküll was able to establish that in the *umwelt* of a trout, time could be broken down into increments of one-fiftieth of a second. This was the trout's moment of visual time, during which it could perceive and distinguish an objective physical sensation of sight.

For human beings this moment of visual time is one-twenty-fourth of a second. It is at this speed that movie projectors operate. Instead of perceiving the *objective* reality of a separate sequence when lantern slides appear on the screen, we get the illusion of a continuity of movement. For a trout, then, roughly two and one-half times as much visual movement information can be crammed into the same moment as could be perceived inside that moment by a human being.

In another experiment he trained a snail to mount a stationary stick for a reward of food. If the snail attempted to mount a vibrating stick, it was punished with shock. He discovered that when the stick vibrated at the rate of four beats to the second, the snail perceived it as being stationary.

In his book, *Theoretical Biology*, von Uexküll wrote: "All psychic processes, feelings, and thoughts are invariably bound to a definite moment and proceed contemporaneously with objective sensations. . . . Time envelops both the subjective and objective worlds in the same way, and unlike space makes no distinction between them." He defined subjective time as being a series of what he called "moment signs"—"the smallest receptacles that, by being filled with various qualities, become converted into moments as they are lived." Intervals of time which do not include the kind of experience that endows them with a quality— either pain, or pleasure, or even simple attention—he called unaccented moments. For the cattle tick, eighteen years may pass as one unaccented moment.

For us, as we age, time appears to accelerate. For a man of middle age, a year seems to pass as swiftly as a month does for a child. Through our exposure to them, accented moment signs —those moments that contain external happenings which mobilize our resources—lose their novelty and demand less attention. We become habituated to them. Such an external happening as the sound of an airplane—which would be a moment-sign to a child, which would mobilize all his senses to integrate and evaluate this external happening—would be subliminal to an adult.   Through habituation to this stimulus we would come to ignore it and the moment that was marked by its happening would be deprived of its accent.   It would become unaccented.

Since von Uexküll worked, around the turn of the century, our understanding of neurology and our laboratory techniques have become more sophisticated.   We now know that linear time is capable of being perceived in different increments by different sense organs, that the rate at which sensations are transmitted is limited in part by the speed at which the nerve synapses transmit electrical impulses.   These speeds vary with differing nerves in different parts of the body.   They also vary grossly from creature to creature.   The brain of a cat is very fast, transmitting impulses at a speed of about 119 meters per second or about 266 miles per hour, while the impulses pass along the neural net of a jellyfish at a speed of about .15 meters per second or about one-third of a mile per hour.

The state of being we call life is a precarious adventure.   A rock in the field merely *is*; it exists in time and undergoes progressive alterations of state without being required to perform any volitional activities aimed at extending any particular phase of these alterations.   But all living creatures—except, perhaps, viruses and other similarly unique organisms occupying the borderline between life and nonlife—possess the quality of life in two distinct phase forms; in the language of genetics they are the haploid and the diploid.   We customarily think of life as being that quality possessed by the diploid phase, but Samuel Butler, in a wonderfully sardonic phrase, jolted this anthropocentric view of life by stating that "the hen is merely the egg's way of producing another egg."   The haploid phase of ourselves is simply the egg and sperm phase.   The term haploid derives from the fact that egg and sperm, when they finally meet, each possess only half the required number of chromosomes.   They fuse, and in that act donate to one another the missing genetic materials.   However, the fact of being haploid can in no way be construed as lacking life.   A very large number of organisms spend the larger part of their lives in the haploid phase—algae and fungi are examples. However, it may not be construed that the haploid phase of life— even human life—is sterile, incapable of producing diploid cells, of forming a fetus.   This phenomenon, the production of diploid from haploid cells, is known as parthenogenesis and takes place quite routinely and normally in several animals.   Perhaps the best known is Daphnia, a fresh-water crustacean commonly found in seasonal puddles of water in temperate zone meadows, marshes, and other similar places.   These animals survive the winter in the form of eggs encased in hard leathery armor. Alternate freezing and thawing in the spring cracks these egg cases and releases the tiny shrimplike animals which are all females.   They produce young, other females, and the colony thrives and increases during the warm summer months without any need of male intervention.   The transition, or metamorphosis, from haploid to diploid is completely asexual.   The haploid phase is therefore in no way incomplete.   It is fully capable of producing adult forms with a total complement of structural and behavioral traits.

It is theoretically possible for human females to be similarly produced by human females, since all eggs, human as well as Daphnia eggs, produce, during the process of maturation, another genetically complete but much diminished copy of themselves known as the Polar Body.   In mammals this polar body is normally expendable; it is thrown off by the egg and becomes absorbed by the tissues of the ovary.   In Daphnia, however, this polar body is reintegrated into the egg, playing the role of sperm and supplying the missing genetic materials.   As fall approaches and the Daphnia colony nears the end of its active locomotor phase of life, males appear.   The present hypothesis holds that the shortened span of daylight causes females to produce them, though their production can be induced by chemical changes in the water as well as alterations of light.   So far as is known, the sole function of males is to produce the hard leathery egg cases which only appear as a result of heterosexual copulation.   The eggs produced without male intervention are soft-shelled and defenseless.

This was Butler's point—the egg merely exists. The hen, moving to and fro and up and down in the world, renders herself accessible to experience. One of these experiences is likely to be lethal unless she avoids it. The commonest way of avoiding lethal experiences is by flight from them.

This then, the act of acquiring time by avoiding death, provides the ultimate source of all anxiety. *Anxiety* is innate, and so, at least in animals, are the ways of dealing with it by means of flight. Obviously, rats, lacking wings, cannot flee into three-dimensional space as pheasants can; rats scurry into their holes. But often flight behavior is not so obviously dictated by structure. The subjective *umwelt* conditions which determine the mode of flight are obscure. There is, for instance, one species of gecko lizard which flees by running upward on a vertical wall, while another, so closely related that it is almost indistinguishable, flees by running downward on the same wall. This flight reaction, distinctly specific to each species of this lizard, has become the easiest means of field identification. Another example: When Raymond Ditmars, the famous herpetologist, wished to capture some vampire bats for the New York Zoological Gardens he found them occupying the same Central American caves as fruit bats. When frightened, the fruit bats took wing and were off like birds. The vampires, however, scuttled away afoot, like rodents, seeking shelter in crevices where Ditmars was easily able to snare them.

The characteristic flight of the tumbler pigeon resulted from inbreeding to obtain a flight reaction that was specific in several species of pigeons. When attacked by falcons in mid-air, these pigeons flip over backward in full flight and make off as fast as they can in the reverse direction. By breeding for this trait, frivolous fanciers were able to produce a breed of pigeon whose inherited memory of flight was so dominated by this one aspect of flight—fear and escape—that eventually all they were capable of doing with their wings was this sterile aerial somersault of survival.

As preparation for flight most animals evacuate the bowel, presumably to make available energy more effective by reducing the weight-power ratio. Von Uexküll was fascinated by what seemed to him an early evolutionary form of this escape-associated behavior in the sea slug. This creature, when affrighted shoots out the entire lower intestine as well as its contents, growing it back anew when conditions permit. In many vertebrates the sudden onset of fear produces a spasmodic diarrhea, and the persistence of this syndrome in man forms the basis of many a hoary military joke. In addition to this extreme manifestation all manner of gastro-intestinal malfunctions, running the entire gamut from simple indigestion to ulcerated stomachs, result from anxiety states in human beings. Though there is no longer necessarily a conscious desire to displace oneself in space by flight, the human body still reacts as if to prepare for it by jettisoning the useless ballast contained in the gut. And the worst physiological damage is done as humans persist in eating while in an anxiety state. The autonomic nervous system prepares to evacuate food in preparation for flight, but the will, obedient to the demands of social custom, crams down ever more where it is not wanted. The peculiar institution of the business lunch is a new behavioral ritual for man. Ruminants have faced this problem of ingesting food while in a state of psychic stress for a far longer period of evolutionary time and have, as a result, evolved a structure to cope with it—the ruminant stomach. The grass-eating ruminant must find its grassy food where grass grows—in open sunlit plains or pastures. In the open, in bright sunlight, it is eminently exposed to predators, and while eating, must be in a continually alerted, excited state, prepared to flee at the slightest warning sign. While thus exposed, the ruminant eats quickly and in large amounts. This food goes right into the rumen, barely chewed. The animal then retreats from the harsh sunlight into the cool protection of dappled shade, whereupon the food is regurgitated from the rumen and is chewed again, this time carefully and thoroughly, so that, in effect, the ruminant does manage to enjoy its food as food should be enjoyed, in peace, quiet, and security.

Not all flight reactions involve a displacement in space. The phrases "frightened stiff" or "paralyzed with fear" refer to sets of behaviors which evolved in order that animals might acquire linear time by doing nothing at all—*akinesis*. The highest, most complex form of this akinetic flight reaction is perhaps best expressed by Hamlet's predicament—the inability to choose between various alternative courses of conduct. Equally highly evolved, but in another direction—that of structural specialization —is the akinetic flight response of many insects, notably the walkingsticks and leaf insects of the order *Phasmatidae*. This response is facilitated and rendered effective by their deceptive appearance. Though relatively common they go unnoticed by all except the most attentive observer, for they look exactly like twigs or leaves. On perceiving the slightest threat, they freeze into inactivity. In Samuel Butler's terms, they acquire time in the fashion of eggs—simply by being, not by movement. They acquire time by disappearing within it, while those animals that acquire time by outright flight, exchange space for time, leaving spatial possession under the dominance of the aggressor, and by its surrender, gaining time.

The means by which animals evolved protective coloration, enabling them to pursue akinetic flight behavior successfully, cannot be understood from a purely economic point of view. The commonplace whippoorwill is an example: A superb flier, it is, presumably, extremely well equipped for escape in space. Yet, more often than not it chooses akinesis as a threat response. The plumage of this nocturnally active bird consists of an irregular pattern of dappled browns and grays. During the day, when it is largely inactive, it can huddle almost invisibly against a background of tree bark. It can perch bellied against the trunk, becoming, insofar as appearance goes, a part of the tree itself.

The coloration of the herring gull can be understood totally in economic terms. In the immature phase it is speckled brown, a blend of colors which render it inconspicuous against a background of pebbled beach. As it grows older and the economic demands of life bring it out over the sea, its adult coloration of gray and blue[1] makes it difficult to see against the background of ocean spume and the leaden skies of northern seacoasts.

The brilliant scarlet of the cardinal serves no camouflage purpose; it is utterly inappropriate to its habitat. This color serves the adult male as a permanent display. Its behavioral rituals in connection with territorial acquisition and mate attraction are obviously more genetically stable than physical structures which evolved in response to behavior. In the subjectively perceived *umwelt* of the cardinal, notoriety was more necessary to existence than prudent camouflage. And the fact of the cardinal's continued existence demonstrates that this is not necessarily unadaptive. The bird has also evolved a shrill, attention-demanding call to augment its appearance.

While the whippoorwill is equipped for akinetic flight responses by coloration, it nonetheless, like the cardinal, has a startlingly beautiful call. The English sparrow, however, is dull in every way, in appearance and in vocalization. Its modest, nearly inaudible chirp serves the very minimal requirements of communication. The Swiss psychiatrist, L. Greppin has written a curious account of an encounter with a population of sparrows which involved mass paralytic hysteria. As director of the Rosegg Sanitorium he was responsible for ridding the institutional premises of hordes of sparrows. After various other attempts to drive off the birds failed, he hired shotgunners to kill them. He himself participated in the hunts. "At first," he writes, "they [the sparrows] only discriminated against me when carrying a gun; then whether armed or not, I was distinguished from the rest of [the medical staff] and at the same time flight distance increased. As soon as I appeared, their cries of fright

---

[1] The official U.S. Navy camouflage tone, "Omega Gray," has the optical properties of the color of the Antarctic Petrel and resulted, in part, from studies made of this bird's plumage.

and alarm grew in number and intensity, and the escape reaction
began still further away.   Then, after about eight or ten weeks,
unusual phenomena of movements and lameness appeared as an
expression of almost panic flight.   The sparrows dropped like
stones into the bushes and then looked as if they had frozen into
rigid postures."

Any attempt to draw parallels to the point of identity between
the behavior of birds and humans is obviously futile.   Millions of
years of accumulated experience have intervened between us and
them.   We are not only different, we are totally other.   The
differences are qualitative as well as quantitative.   Yet the whole
point of ethology lies in its attempt to contrive a theoretical
framework by which behavior can be comprehended comparatively in the same way that anatomy is comprehended
comparatively.   By the end of the last century, if a skilled comparative anatomist were given a scrap of fossil bone, he could
construct, by analogy, its place in the skeleton of an extinct
creature and possibly even make reasonable deductions concerning the classification of this creature.   There is no longer any
doubt whatever that among existing vertebrates anatomical
structures are homologous, and that those two bones—the radius
and the ulna—are homologous in the wing of a bird, the foreleg
of a dog, and the arm of a human.   The whole science of ethology
is committed to the accumulation of sufficient behavioral data
from all animal phyla, so that theoretical models of the evolution
of behavior may be constructed.

Within this framework Greppin's paper would represent
anecdote only.   A modern ethologist would want to know a
great deal more about these sparrows, acquiring a great many
more statistical data before making any judgments.   But if one
grants that Greppin, on the basis of his scientific training, was a
competent observer, and that his report is basically true in its
general outlines, then further puzzles intrude.   In this instance
the sparrows, responding as they did to this threat of time's
curtailment, behaved (in psychoanalytic terms) regressively.

They reverted to behavior appropriate to a former, immature phase of life. Had they been nestlings, akinesis would be a perfectly appropriate adaptive response to threat.

Hysterical paralysis, also in response to threat, is not unknown among human beings. However this behavior pattern is never innately appropriate in the life history of humans. We have no structural accommodations for making akinesis adaptive; we possess no protective coloration. When confronted by a threat to life—a threat to time's curtailment—human infants respond by crying, kicking, and generally clamoring for attention. This is in marked contrast to the akinetic response to threat of many infant animals. If hysterical paralysis in humans is to be considered regressive at all, it cannot be considered an ontogenetic repression (a return to a former state of being in the personal history of the individual), but a phylogenetic regression (a behavioral casting backward into the collective history of the species). At some point in our primate pasts akinesis as a threat response must have been appropriate, and though our structures have altered in the intervening time, remnants of this response continue to persist.

Behavioral components are not nearly so easily measured in comparative terms as are skeletal components. One can compare two bones one with the other, minutely as well as generally, using calipers or microscopes or any other device. But behaviors are more transient in time than are objects. For study purposes they can be recorded on film, and examined in slow motion or stop motion, and when funds permit the use of this expensive technique, modern ethologists avail themselves of it. But a strong bias still runs against this type of study; it has to do with the nature of reality as existing in time. Behavior is transient. No one questions the existence of a rose. As Gertrude Stein wisely remarked: a rose is a rose is a rose—and that's all. But it is very hard for many scientists to consider behavior existentially, as they would a rose.

One attempt to do so is to consider behavior in terms of roles.

It has become fashionable lately to consider many of mankind's interactions with himself as games; and since a game has only a limited purpose, since it has no impressive teleological implications, it is a useful device for maintaining objectivity. We do occupy certain positions at any given moment in relation to other people and their positions. Costumes are an important attribute of these games. The soldier and the policeman wear special occupational costumes and play specific social roles while wearing them. So do we all. We wear business costumes for business games, sports costumes for recreational games, and no costumes at all for carnal games.

The whole dilemma of games and costumes is encapsulated within the word *habit*: and the difference of superficial meaning between say, *riding habit* and *drug habit*. Originally the word *habit* derived from the Latin *habitus* meaning dress or appearance. But in current usage it is applied to a set of roles that one plays in relation to oneself. The essential nature of role-playing as an anxiety suppressor is obvious in the habits of certain neurotics who are compulsively clean, or neat, or must avoid, at the cost of convenience, walking on the cracks in sidewalks, etc.

Once removed from its purely economic-determinist matrix, the process of natural selection can be seen as promoting adaptation within a specialized role. Particularly among insects, where social rank and role is almost totally determined by structure and appearance, structural aids to role-playing abound. There are innumerable examples of the meek and mild prey insects, which masquerade in the mimicked appearance of fiercer predatory creatures. But role-playing—the putting on of extrinsic, non-structural costumes in order to alter appearance—is comparatively rare. Possibly the most famous example is found in the crustaceans, among the five genera of so-called masking crabs. They possess structural aids to masquerade—their carapace, or upper shell, is covered with sparse rows of barbed bristles, and their pincers have evolved in such a way as to allow them to reach anywhere upon this carapace. Adolph Portmann, the eminent

Swiss ethologist, has described their behavior: "The crab uses its pincers for picking off small algae and cutting sponges to exactly the required size. The crab then puts these into its mouth and chews them for a few seconds before rubbing them firmly on its body. Should any of the substances fall off, they are popped back into the mouth and chewed again. No doubt they are covered with a sticky secretion in the process of chewing, for the crab neither swallows nor changes the shape of the particles. If this 'mask' is pulled off the crab, it becomes very restless and rushes about in search of a new disguise. Each kind of crab has its own masking method: thus *Maia verrucosa* always starts with the head and very often leaves it at that; and *Posidonia* lengthens the head, and thus the whole body, by applying strips of material; while *Inachus scorpio* only camouflages its first pair of very long walking legs, which are then inclined at any angle to the other legs.

"An entirely different kind of disguise is used by the sponge crab *Dromius vulagaris*. This crab, after cutting out its colorful spongemask, holds it in place over its back by a specifically modified last pair of legs, which are shorter and carried up higher than the other ones. The sponge crab is an expert cutter, for if we offer it a large piece of wet paper, it will immediately cut out a piece of just the right size. If its paper coat should fall off, it will quickly design a fresh one."[1]

So far as is currently known, no mammal, except the two-toed

[1] Though its role is an aggressive one, and its masquerade largely structural, the role-playing life of *Aspidontus taeniatus* is curious. Groupers and other large fish inhabiting the warm clear waters of the Caribbean employ cleaning fish in the same manner as rhinoceroses and crocodiles employ tick birds to cleanse their skin and teeth of parasites and debris. The neon goby is such a cleaning fish. Much smaller than its hosts with whom it has evolved this symbiotic relationship, it is brightly colored and swift-moving. Groupers open their mouths to the neon goby, giving them access to their teeth; they lie motionless for the neon goby to clean algae and other matter from their gill openings. *Aspidontus* is a ferocious raptore, but so small it could not successfully attack a grouper by frontal assault. It has evolved the appearance and manners of the neon goby with the result that groupers mistake it for the cleaning fish and allow it a close approach expecting the usual treatment. Instead *Aspidontus* tears out a few large chunks of grouper flesh and is off before the larger fish can react with a suitable defense.

sloth and man, employs extrinsic materials on its body for time acquisition purposes. This large, slow-moving, defenseless animal, inhabiting South American jungles, allows green algae to take root in its fur, giving its coat a gray-green cast which blends nicely into the misty shadows of its rain-forest habitat.

On primate behavior, of course, we have more complete data, and from what we know, the wearing of human clothes may well have evolved, not exclusively as a response to the need of warmth nor as time-acquiring masquerade, but for role and rank delineation. During the First World War, Wolfgang Koehler, one of the founders of Gestalt psychology, was interned on Tenerife in the Canary Islands where the German Government maintained a biological station. In this station he kept a colony of chimpanzees, allowing them considerable latitude and freedom. He wrote that they picked up scraps of burlap which they found lying about the station and "spontaneously" decked themselves out in this cloth. More recently Miss Jane Goodall, studying wild chimpanzees in the Gombe Stream Reserve near Lake Tanganyika, reported similar thefts of cloth: ". . . old clothes, greasy cloths from the kitchen were the most sought after," she writes, and "David [one of the apes who, when he found out he could enter the camp unmolested, became quite tame] went off with a good many blankets as well as shirts and other garments, and Goliath took many tea towels. . . ." Goodall however, does not mention seeing the apes dressed up in these borrowed clothes. It is possible that these antics of the chimpanzee, either in the wild or in captivity, are mimetic; monkey-see monkey-do. However, it is interesting to speculate on the idea that human clothing may not have made its appearance in response to the need for body warmth but rather in response to role and rank delineation. In the instances reported by both Koehler and Goodall, the chimpanzees were "outranked" by the humans with whom they gradually came into close and almost intimate contact, but of whom they were nevertheless fearful. It is possible that the chimpanzees stole clothing in order to "buy" status.

[2]

Cyclical Time

In his experiments with the trout and the snail von Uexküll was concerned with the perception of elapsed, or linear time. The accented and unaccented moments are increment measurements, and the time perception of the trout, being superior to that of the snail, was superior in the same way as the steel rule of a machinist is superior to the wooden ruler of a schoolboy. It measures gradations in smaller increments and more accurately.

All prey-predator relations involve competitions in linear time: the kingfisher must be able to cut time into even smaller pieces than the trout in order to catch it in mid-passage. The mongoose must be able to move in for the kill faster than the cobra can recover from its strike. The cheetah must be able to run down the antelope, and so it goes. When we humans think about time, we usually think of it in those linear competitive terms which are the source of anxiety. Regardless of its duration, life is still shortened by each passing moment. Elapsed time will never come again, and the great anxiety of the West is to cram each moment full of—not necessarily perception—but accomplishment.

By far the largest number of our technological achievements consist of devices which manipulate time in terms of something else—except of course for modern medicine which, by extending the life-span, quite simply and directly draws out biological time in its own terms. But our achievements in communications and transportation—jet planes, ballistic missiles, radio and telephone, and soon—compress space in terms of linear time, thus expanding the latter.   By far the largest number of machines using extrinsic power sources compress labor in terms of linear time, once again extending linear time, permitting us leisure.   And that, according to the dictionary definition, is only "unengaged time"—time for living, pure biological time.

However, seen purely as a biological phenomenon—that is taking time *out* of the environment, not making it extrinsic to the organism, but including it as part of the stuff of life itself—the linear aspect of time is its most frivolous aspect.   In biological systems time represents the metabolic process, the absorption and utilization of energy.   And seen from this point of view, time is rhythmic—the heart beats, the respiration goes in and out; time is cyclical.

This is the aspect of time most important to all animals and humans except those of us tightly caught in the coils of a technological society.   For primitive hunter-gatherer man, or even for that small section of our modern population involved in settled agriculture, the most important view of time is the cyclical one, time as measured by the passage of the seasons, the coming and going of equinoxes and solstices, though their power and influence varies with the latitude.   The effects of the alternation of summer and winter, of spring and fall, are less extreme at the equator and at the poles than they are in the temperate zones. This succession of the seasons provides us with the great cycle we call the year.   The next most important biological cycle would appear to be the twenty-eight day lunar month.   Though we moderns reckon our calendars by solar months, that is really only a bookkeeping convenience. Many animal behaviors are dominated by the phases of the moon, by the cycles of its waxing and waning, and more and more human behaviors and biological processes are now considered to be affected either directly or indirectly by lunar periods.   Of course the smallest and most dramatic cycle which dominates our behavior is the diurnal period, the rotation of the earth, the alternation of night and day.

Yet we don't live our lives in accordance with this knowledge of the importance of cycles on our physiological and psychological well-being. We know that flowers bloom in the spring, and that the swallows *return* to Capistrano, but as regards the conduct and apprehension of ourselves inside the phenomenon of time, we are still entranced by the fallacy of Western causal logic. The fallacy involves the idea of closed systems. There are no closed systems in nature; everything involves everything else.

The great reward of Western causal logic has been technology and the manipulation of our environment. The loss has come about because we consider every act a closed system with short range predictable consequences. The result is therefore a loss of meaning to the act.

If a Westerner heats a kettle of water over a flame of natural gas, he can either brew a cup of tea, or divert the steam into a piston and use it as a power source. At any rate the steam results from the heat and according to the nineteenth-century physical view of things, the heat, the water, and the steam were a closed system.

In the nonindustrialized Orient, the Hindu notion of Karma still prevails. This is an expanded notion of causality. According to Karma every act reverberates forever throughout the universe as a whole, in time past as well as time future. The natural gas, which is burned to provide the heat, was once a living entity, very likely a fern (alive 200-odd million years ago in the Paleozoic era), which underwent several transformations, or changes of state, or incarnations, finally appearing in time-present as combustible coal. The implications of its burning extend into time-future, inasmuch as the total supply of coal is now depleted by that amount used to heat the water. Future generations shall be poorer in coal as this amount is transformed into still another state of being, light and heat. The man who heats the water is transformed to some extent, either by drinking the tea or using the steam power to perform a task. He is not what he would have been had he not performed the act. And so it goes, in dizzying gyres, paralyzing the will. If every act involves such awesome consequences, and one is responsible to the past as well as the future, one is lost in contemplation and hesitates to act in the Western sense. We tend to become impatient with the apparent passivity and inertia of Oriental ways when they involve our interests.

Yet crucial to this system of Hindu thought is the notion of cyclical time. Modern biologists are coming more and more to realize that by closing our systems, confining them in terms of linear time, we have narrowed our understanding of ourselves and have, moreover, done ourselves biological injury. The idea of Karma also involves another aspect of time, the notion of the ever-moving present, a commonplace idea among primitives but so totally foreign to causal logic that we Western sophisticates find it difficult to reconstruct our thinking to comprehend it. According to Karmic logic, time moves in layers, like a glacier, and the bedrock at the bottom is the present, was the present, and always will be the present. Only the superficial phenomena, the junk picked up and pushed along by the glacier itself, moves and is changed in time by abrasion with the bedrock, the ever-moving present. And the present is in fact ever-moving, because the universe as a whole is aging and becoming something other than what it is. The Platonic notion that everything knowable is known, that learning consists of recollection, rests on the assumption of the ever-moving present. It seems to me that we can only come to comprehend some of the wondrous oddities of human and animal behavior when we examine them *outside* of phenomenological time.

It is a commonplace for all of us these days to make the attempt at self-understanding through apparent causal chains conceived of as occurring within a linear time scheme. For example: a man has a "bad day," a day of fatigue and depression and generally lowered efficiency. If this day follows upon a sleepless night of worry, he finds it easy and natural to create a cause-and-effect relationship between these two happenings, attributing the effects

of the day to the cause of the previous night.   If this man, being
of introspective temper, can then trace the prime mover in this
sequence to a so-called anxiety-producing occurrence which
caused him, on the day previous to the sleepless night, to drink
too many cups of coffee and smoke too many cigarettes, he can
then stretch the causal chain still further backwards in linear time.
And should he so choose, by attaching a Freudian causal system
onto this chain, he can theoretically represent the origin of the
anxiety-producing occurrence as dating, possibly, from the first
three years of life.

The principal difference between this kind of Western psychological construction and the Hindu notion derives from the fact
that the Western system operates solely within a linear time
scheme, while the Karmic notion involves the idea of cycles.
Another way of visualizing Karmic consequence is to imagine
the bow wave of a boat.   Seen from a distance, the wave appears
almost like an attribute of the boat; it is a constant function of the
boat's speed.   Yet seen close by, it is continually recreating itself,
and if one watches a single fleck of foam that *is* the wave, one sees
that no wave ever repeats its action.   The wave is a cyclic event,
and the individual particles within the wave do not, by themselves,
move in the form of the wave.   They move perpendicularly
to the wave form.   They move up and down in an elliptical path,
like a bouncing ball with ever-diminishing amplitude.

If our man were to examine this bad day from another point
of view, taking into account the phase of the cycle, he might
consider those events which occurred adjacent to it in time—the
nervous anxiety, the excessive coffee-drinking and cigarette-
smoking—to be coherently rather than causally related.   For
example, we know that there is no causal connection between a
sore throat and a stuffed nose even though they normally follow
one another in that order.   We also know now that neither of
these symptoms of the common cold is causally related to a cool
draft of air.   The *cause* of the common cold is a virus.   In the
same way we are just beginning to realize the overwhelming

importance of cyclical time as a determinant of body-mind states of being. Our mechanically linked chains of cause-and-effect are incomplete. We must consider cyclical time as a factor.

It is curious that we are again, just at the present moment in the intellectual history of the West, becoming interested anew in cyclic events. Prior to the Enlightenment, we were far more open minded about cyclic events and their relation to behavior (or destiny)—not only about animal cycles, but human cycles as well. A knowledge of the lore of cycles was an indispensable attribute of learning, and men learned in planetary and seasonal cycles were consulted on every occasion of consequence by individuals and by the group as a whole.

Since human behavior is more plastic and malleable than the behavior of plants and animals, since man can adjust himself to changes in temperature, altitude, humidity, etc., he can also adjust to minor changes in his cycles. The effects of cyclic time on behavior are extremely subtle. As technology began to effect dramatic manipulations of linear time, the shadowy influence of the cycle was overlooked. But recently—as jet travel has become commonplace, and people have found themselves transported from one hemisphere of this planet to the other in a matter of hours, thus disrupting the individual's relationship to the positions of the sun and moon drastically and suddenly—modern medicine has rediscovered the fact that cycles do affect behavior most markedly.

Since all living systems depend in one way or another upon our nearest star, the sun, for energy, the most universal metabolic beat is the alternation of light and dark, the diurnal rhythm. During the course of evolution organisms have evolved complex variations on this basic rhythm, so the term *diurnal* no longer suffices to describe it. Dr. Franz Halberg of the University of Minnesota Medical School in 1959 coined a new term for this metabolic beat, and his name for it—*circadian rhythm*—has since received wide acceptance. It is a compound word formed of two Latin roots, *circa* meaning about, and *dies* meaning day. The cycle of time in which all living systems exist conforms roughly to a twenty-four hour period, even though this cycle may not necessarily be clued to the alternation of light and dark. In the same way a woman's menstrual cycle conforms to the lunar period of twenty-eight days, though it need not be triggered by any particular phase of the moon.

Probably the first modern scientific investigations into circadian rhythms were conducted by a French astronomer, de Mairan, in 1729. He was puzzled by the observation that the circadian rhythms of plants seemed to be endogenous, that is the plants somehow contained within themselves a sense of cyclical time. De Mairan kept some plants in a darkroom in constant temperature and noted that the sleep and waking movements of their leaves persisted even though no environmental clues were presented them. Thirty years later another botanist, suspecting that perhaps de Mairan's darkroom was not totally lightproof, took plants onto a cave and noted the same phenomenon; that the sleeping-waking rhythms persisted in a regular way under conditions of total darkness and constant temperature. Ever since that time botanists have been actively studying circadian rhythms, and they've discovered that rhythms can be altered, somewhat by speeding up the environmental cues of alternating light and dark (the jargon word for these environmental cues is now *zeitgebers* from the German, literally meaning time-givers) the rhythms can be accelerated. Obviously, if rhythms can be accelerated, growth can be hastened, and agricultural yields improved. With pragmatic commercial advantages in the offing, botanists suffered no shortage of funds to pursue their circadian rhythm studies. They soon discovered, however, that the cycle was relatively immune to gross tampering. It could be speeded up to occur within a span of eighteen hours, or slowed to a period of twenty-seven hours, but these seemed to be the limits.

The botanists' interest in cyclical time quickly spread to other biological disciplines. In 1961 an international symposium was convened in Cold Spring Harbor, New York, to discuss the latest work in biological rhythms. Erwin Bünning, Director of the Botanical Institute at the University of Tubingen, Germany, delivered the opening address. He spoke of the importance of circadian rhythm studies to human health and happiness in the following words: ". . . thus it is possible under certain conditions for the different organs to cease to pay attention to one another. For example, a glandular tissue may be in the phase of hormone production while another organ, being in another phase, cannot make use of the hormone; or an enzyme may be very active in a particular tissue at a time when its substrate is not available. Every transatlantic air traveler knows the physiological discomforts that may arise from such a lack of co-operation. . . . Not only air and space travelers, but all of us in this age live under conditions which can easily induce such unbalanced phases of diurnal oscillations in our individual organs."

The tremendous contribution that botany has made to circadian rhythm studies derives from the structure of plants themselves. They have no brain, no central nervous system, yet they live their lives in conformity to cycles of time. They display *behavior*, they become active during the day, they sleep during the night, they bloom, and they fade. And because this activity is not directed by any system of specialized cells like a brain, or a central nervous system, the botanical approach to circadian rhythms focused on the cell itself, and on the community of cells which comprise the entire organism.

In man as well, the feeling persists that this sense of cyclical time may not reside in the brain, nor in the central nervous system, but in all the cells of the body as they represent its parts. In the First Corinthians, St. Paul wrote about this subject: "The body is not one member but many" and he defined the healthy man as one in whom "there is no schism in the body but that all the members should have the same care for one another."

The growth or development process in man—the very organization and transformation of an individual from a microscopic egg into first a recognizable mammalian fetus, then into a child, and then a man—exists in time (as do all processes) but it is *ordered* by time, and there seems to be no conscious sense on the part of the fetus, the child, or the adolescent that any particular linear moment of time is important. The intellect does not intervene with a sense of time, as it does when we feel it is time to get up in the morning, or time to eat, or time to retire at night. Just because this ordering flow of time takes place *outside* consciousness, we have tended, until just recently, to ignore it.

The diurnal rhythms of our earthly time derive from the earth's rotation. Since this is so, rapid displacement from one part of the planet to another, particularly from one hemisphere to another, disrupts the orderly flow of time. Space and aviation medical researchers have become extremely interested lately in circadian rhythms, and the way in which humans adapt (or *entrain*) to new rhythms, in response to new locations.

But one of the most fascinating and dramatic experiments on this subject was conducted by a zoologist, Dr. Frank A. Brown, Jr., of the University of Illinois, at Evanston. He chose not a man, nor a monkey, nor even a mammal, but the ordinary fiddler crab. This creature displays its sense of cyclical phase by a marvelously distinct signal. It changes color. At high tide the carapace of the crab is yellow colored—a pale yellow, almost white. But as the tide recedes, melanin secreted in special cells of the carapace becomes dispersed and the shell darkens to a dark greenish gunmetal color. We are still ignorant of the adaptive purpose of this change. There are several theories: that perhaps it serves to regulate temperature in this cold-blooded animal, or that it may serve as a shield to ward off harmful ultraviolet radiation. No one yet knows for certain. It is a most unlikely camouflage protection, since at its darkest the crab scuttles across brilliant sunlit sands picking over the flotsam left by the tide, and it would seem that its dark color would make it an irresistible target for predatory gulls. Each day the crab reaches its peak of darkness at the point of lowest tide, and each day this point of maximum darkness is reached fifty minutes later than the day before in strict accordance with the fifty-minute delay of the moon's orbit around the earth, as compared with the earth's turning in relation to the sun.

Brown captured some fiddler crabs one summer on Cape Cod and brought them back to his laboratory in Evanston, keeping them in shallow tanks where they could not feel the oceanic pull of the tide—in a darkroom where they could not see either the sun or the moon.

In the beginning they kept their color changes clued to the tides of faraway Cape Cod, but gradually over a period of several weeks their sense of the moon's passage through space caused them to alter their rhythm until at last they were changing to their darkest color when the moon was directly overhead. The moon pulls at the atmosphere, the ocean of air surrounding the earth, as well as it does at the oceans of water, and the effect of this pull can be noted as changes in atmospheric pressure. To discover whether barometric pressure was the indirect factor which tied the crabs to the moon's circuit, Brown installed his crabs in double tanks—a tank within a tank—and arranged through a hookup to a barometer for the inner tank to be raised or lowered in accord with alterations in barometric pressure. New crabs were caught at Cape Cod and flown to Illinois and installed in these double tanks. Brown waited to see whether they would now sense that they'd been relocated on the planet, where the moon would pass over them at a new time, and whether they would change their color accordingly. They did, and Brown writes, "To our astonishment the correlations still persisted even though the organisms themselves were now held in unvarying ambient pressure. And the correlations were of such character as not to be expected by chance. Clearly, some token stimulus reflecting outdoor pressure changes was still reaching the organisms. Later we learned that information reflecting atmospheric temperature was also reaching the organisms though their ambient temperature, similarly, was meticulously held constant. Underlying more or less 'noise,' both outdoor temperature and pressure are rhythmic with all the solar and lunar periods."

But the period itself, its rhythm, the duration and "shape" of the cycle is jolted out of its sequence if the animal moves across the surface of the planet. It seems now in retrospect that any naturalist would have realized long ago that by referring to time, to the position of the planets, the moon and the sun, animals might most easily orient themselves in space. Unlike human beings, most animals inhabit either a barren or unstable landscape —to fish in the sea, birds migrating high in the nighttime sky the landscape is barren. For all practical purposes they inhabit "pure" space, empty space. Those euclidean points of reference by which we large mammals ordinarily locate ourselves are not available to them. Many insects confront a similar problem in orienting themselves; their landscape is unstable—a puff of wind, the footprint of a larger animal, can utterly destroy the familiarity, and therefore, the usefulness of landmarks. Some stable reference system which these creatures could employ must be imagined. And what more stable system than time, that innate sense of planetary cycles with which all living organisms were provided from the beginning? Yet it has been so long since man made use of such a system innately, that it has quite literally slipped his mind. Seamen and aviators, when confronted with this same kind of barren landscape, must resort to a similar system of time orientation in space, but since this kind of time consciousness is no longer innate, human navigators must be provided with elaborate star charts, moon and sun tables, sextants, chronometers, and all manner of paraphernalia so that reason can operate in those areas where "instinct" has atrophied.

Once again human vanity has blinded us to the possibility that these accomplishments need not necessarily result from the intellect. Strangely enough, August Forel, the Swiss psychiatrist and sex-reformer (who, like Kinsey, was also an entomologist), first published observations around the turn of the century, suggesting that animal time sense might be more sophisticated and hitherto suspected. Forel took his breakfast outside in his garden during the summer months. He noted that honeybees came to taste his marmalade at the appropriate hour each morning *even when no marmalade was present* on the breakfast table. They even buzzed the area when no table was present. Forel concluded that, unlike flies, honeybees were attracted to the marmalade by something more complex than a simple stimulus response chain, causally related. He published his observations and there the matter lay for some forty years. However, ten years later a young Viennese zoology student at the Zoological Institute at Rostock (where von Uexküll had been studying the time sense of the cattle tick), named Karl von Frisch, became interested in honeybees. Von Frisch's curiosity had been piqued by a series of papers published in 1910 by an ophthalmologist, C. von Hess, who claimed that honeybees lacked all sense of color discrimination. Von Hess had projected light through colored filters and failed to get recognition responses from honeybees. Von Frisch was interested in visual phenomena and suspected that the difficulty with von Hess's experiments had to do with projected light. In order to test the color sense of honeybees, von Frisch began training wild bees to select small bowls of sugar water from other identical bowls of ordinary water on the basis of the color of a paper doilie on which the bowl was placed. If the bees learned to associated the color blue with sugar water, they would ignore bowls placed on yellow or red doilies, flying straight to the bowl situated on blue paper, regardless of how the bowls were spatially arranged in relation to one another. Von Frisch proved that, contrary to von Hess's contention, honeybees had a well-developed sense of color discrimination, and later experiments proved that they also possessed a sense of form; that they would select sugar water inside boxes marked with various forms like crosses, circles, and stars, etc. But his most far-reaching discovery had nothing to do with color or form—it had to do with such human traits as communication, and linguistics, and time-space orientation.

This discovery came by accident. "When I wish," he wrote, "to attract some bees for training experiments, I usually place upon a small table several sheets of paper which have been smeared with honey. Then I am often obliged to wait for many hours, sometimes for several days, until finally a bee discovers the feeding place. But as soon as one bee has found the honey, more will appear within a short time—perhaps as many as several hundred. They have all come from the same hive as the first forager; evidently this bee must have announced its discovery at home." How was this knowledge communicated?

Von Frisch first located the hive. Then he made arrangements to identify each individual bee, so as to determine whether there were communicators of information and receivers of information, or whether all bees in the hive were equally able to give out and receive information. He devised a two-digit color code for the abdomen and a single-digit color code for the thorax. He suspended pigments in shellac and placed dots of color on each animal until he could identify each of 999 bees in a hive.

He then moved the colony into an artificial hive with glass walls, so he could observe the activity within. He noted that when a bee discovers a new source of food, it returns to the hive and "begins to perform what I have called a round dance. On the same spot, he turns around, once to the right, once to the left, repeating these circles again and again with great vigor. Often the dance continues for half a minute or longer at the same spot." The dancer may then move to another part of the hive and repeat the dance.

Von Frisch also noted another dance, which he called "the tail-wagging dance." In this performance the bees "run a short distance in a straight line while wagging the abdomen very rapidly from side to side; then they make a complete 360° turn to the left, run straight ahead once more, turn to the right and repeat this pattern over and over again."

The deciphering of the coded message of these dances became frustratingly difficult. After von Frisch finally succeeded in breaking them, in the late 1940's, they seemed simple enough. But the initial difficulties centered around the fact that human language is digital, while animal communication is analog. Language is symbolic, and though we humans make analogies (slippery as an eel, strong as an ox, etc.), the analogies are rarely built into the structure of language. For example, if I say I am sad, or I am angry, the precise degree of sadness or anger is not known at this point. Even the addition of an adverb such as *very* angry, or *somewhat* sad is not enough. To communicate the degree, the exact point upon a sliding scale, we must resort, like animals, to analog communication devices—tones of the voice, facial expressions, gestures, postures, etc. In mathematical terms a slide rule is an analog computer, whereas an adding machine is a digital computer. When using the slide rule, the analogy or relationship of one scale to another is immediately and directly visible, while in an adding machine digits must be mechanically piled on top of one another, or removed from the pile, and no *relationships* are apparent.

Since conventional code-breaking techniques were devised for deciphering digital symbolic communications, von Frisch was compelled to begin at the very beginning, trying to find the *analogy* within the pantomime of the bee's dance and the location of the food source. It took him many years of frustrating, painstaking effort.

Now that we know the key, the problem seems childishly simple. But it only became apparent to von Frisch gradually, as he began moving his food source progressively farther from the hive. After the food was moved more than fifty meters from the hive, the bees stopped indicating its direction by means of the round dance, and transferred to the tail-wagging dance. Von Frisch found there were correlations between the duration of the dance and the flying time needed to reach the source. Tail winds or headwinds would alter the length of the performance even though the food source was equidistant in terms of meters.

At last it finally happened—the connection was at last made

between directional space and circadian rhythms. Von Frisch writes of this momentous discovery quite matter-of-factly: "When we watched the dances over a period of several hours, always supplying sugar at the same feeding place, we saw the direction of the straight part of the dances was not constant, but gradually shifted so that it was always quite different in the afternoon from what it had been in the morning. More detailed observations showed that the *direction of the dances always changed by approximately the same angle as the earth's rotation and the apparent motion of the sun across the sky.*"

Unfortunately von Frisch gives us no more intimate glimpse into his personal thought processes. How did he make this fantastic connection between the direction of the bee's dance and the movement of the sun? Was there perhaps a shadow cast upon the hive by some projection which operated like a sundial, and did he notice the connection between the shift of this shadow and the direction of the bee's dance?

The next step was simple enough. The movement of the sun was regular but nonetheless unstable. If the bee were to indicate direction by analogy to the position of the sun, there would have to be some stable convention by which the analogy could be presented in dance form. Von Frisch noted that the dance always took place upon a *vertical* surface. So the next great leap into the breaking of the code came when von Frisch hypothesized that within the hive, in darkness, the one stable directional indicator might be the force of gravity. Working from this assumption, he finally succeeded in breaking the code completely. "If," he writes, "the run points straight down, it means 'Fly away from the sun to reach the food.' If during the straight portion of the dance the bee heads 60° to the left of vertical, then the feeding place is situated 60° to the left of the sun."

However, there was another complication. Even on cloudy days, when the position of the sun was not apparent to the bees, they continued to communicate the location of food sources by the analogy of the dance. His first thought was that the bees were aware of the azimuth of the sun even when it was invisible (in much the same way that Brown's crabs were aware of the position of the moon). He tested this assumption by providing artificial sunbeams—by moving the experimental hive into the shade and diverting sunbeams from the wrong direction with a mirror—and discovered that by doing this, he could disorient the bees. Yet artificial light sources, such as a powerful flashlight beam, would not disorient them.

At this point, in the early 1940's, von Frisch's work could have taken two directions: he could have explored the mystery of circadian rhythms, the mechanics of the innate sense of time—for the bees would necessarily have to be aware of the regular movement of the sun in order for them to utilize this movement as a navigational aid—or in conformity with the principle of Occam's razor, he could assume the more simple explanation, that no complex time sense need exist innately within the bee, that merely some optical property of sunlight made the position of the sun visible to the bee when it was shrouded from human eyes by cloud cover.

Von Frisch chose the latter course. Optics had entranced him from his youth; he had begun working with honeybees in the first place to disprove von Hess's contention that they were color-blind. His thirty-year involvement with breaking the code of their dance was kind of a side trip from his major goal. Perhaps he was happy to find a new optical problem to solve. At any rate, he writes: "Light rays coming directly from the sun consist of vibrations that occur in all directions perpendicular to the line along which sunlight travels. But the light of the blue sky has not reached us directly from the sun; it has first been scattered from particles in the atmosphere. This diffuse, scattered light coming from the sky is *partially polarized*, by which we mean that more of it is vibrating in one direction than in others."

In experiments which he conducted in 1948, he used a sheet of polaroid plastic "about six inches wide and twelve inches long, and every part of the sheet acted as an analyzer; the light passing through it became polarized in one direction, I placed this sheet over the glass window of my observation hive . . . and it became clear at once that the dance was markedly affected." With this discovery—that the bee is capable of discerning the plane of polarized light—von Frisch seemingly lost all personal interest in the mystery of circadian rhythms. He entered the dissecting room and began studying the lense and retina and neural structure of the honeybee's eye. It was a colleague of his, Martin Lindauer, who picked up the track of circadian rhythms.

Lindauer found that he could induce honeybees to *dance at night*, when there was no illumination whatever, no polarized planes of light to suggest the direction of the sun. Once again it was a simple, perhaps even an accidental occurrence that produced this discovery and reopened Pandora's box of circadian rhythms. Lindauer once put a small bowl of sugar water inside the door of the hive. "Then," he writes, "the bees start to dance, but do not announce that there is food nearby, namely in the hive. Instead they remember the last customary feeding place outside the hive and carry out a tail-wagging dance which indicates the direction and distance of this goal. It is very astonishing that these dancers indicate the correct sun azimuth at night, although, of course, they have never seen the nocturnal position of the sun."

So there were now two mechanisms by which the honeybee oriented itself in relation to the apparent position of the sun: one was the plane of polarized light, the other was what appeared to be an innate knowledge of the sun's predicted orbit! While it might be difficult to prove the latter contention, it would certainly be simple enough to prove the validity of Forel's contention that bees "keep appointments" with food. In order to test this hypothesis, Lindauer placed a bowl of sugar water to the south of the hive in the morning, then removed it after a while, and in the afternoon placed another bowl of sugar water to the east of the hive. The bees kept their appointments with the food. This was not unexpected. But then Lindauer did something else; he fed the bees inside their dark hive at night, thus inducing them to dance and indicate the direction of an *external* food source. Which food location would they indicate by their dance at which time? Lindauer writes: "Position and feeding time are memorized in the dark hive at any time. If now the bees are induced to dance during the night, the choice of dancing direction is not at random. . . . In one case, for instance, a dancer announced at 9:30 P.M. a feeding place in the east at which it had been fed every day at 6 P.M. At 3:45 A.M., however, the dancer announced a feeding place in the south at which it had been fed every day at 8 A.M."

In this instance the bees not only remembered in the late evening, after the sun had descended, where they had last found food, but much more astonishingly, they were able to anticipate in the early morning hours, before the sun had arisen, where they could next expect to find it in time future.

Now, if we human beings are awakened from our sleep at night, it is difficult for us to know the time; whether it is midnight, or three in the morning, or just before dawn. This is not so during the day. We can roughly gauge time by the position of the sun; we can tell generally whether it is morning, noon, or afternoon. Does the honeybee sense the passage of time innately, or does it respond to clues provided by the environment? Since the dancing takes place inside the dark hive, it appeared unlikely to Lindauer and to his colleague, Max Renner, that any clue which could easily be imagined by humans could penetrate the darkness. It began to occur to Lindauer and Renner that by transporting the bees to a completely different time zone they could determine whether the time sense was innate or provided by a *zeitgeber*. Max Renner writes: "In 1955 I had the opportunity to carry out such an experiment [to prove] that the time-sense of bees is able to function independently of diurnal exogenous factors. Bees were trained in Paris for several days to collect sugar water from 8:15 to 10:15 French Daylight Time. After training, they were flown overnight to New York.

At the test in New York, they came to the feeding place at about 3 P.M. Eastern Daylight Time, that is, twenty-four hours after the last feeding in Paris.   No bees appeared between 8:15 and 10:15 Eastern Daylight Time.   The reciprocal experiment (from New York to Paris) had analogous results." This was conclusive proof that the time sense was innate, that it did not depend on a clue provided by the environment.   Had there been an environmental clue, a trigger, a stimulus of some sort, it would have overridden the time change induced by the transatlantic voyage.

Every sense possessed by an animal seems sooner or later to be transformed by the necessities of survival into the service of a skill. In the last fifty years zoologists have come to notice particularly that, in environments where light is missing and therefore the sense of sight no longer functions, animals that are required to orient themselves in dark spaces effectively employ their sense of hearing instead of sight.   Bats and certain cave-dwelling birds have come to use an echo sounding system, uttering rapid squeaks or clicking sounds, then listening for the echo pulse to determine whether they are approaching an obstruction.   We humans do the same thing; the tapping of a blind man's cane is also an echo sounding device.   By long attentive practice, the blind man can tell from the sounds of the tapping what lies ahead of him.

The operative factor in using some other sense than the customary one to fill a sensory gap in perceiving the *umwelt* seems to be a pressing need that such a skill be acquired.   The missing or unknown part of the *umwelt* must be disclosed.   If the bat or the blind man were content with immobility, there would be no need for echo sounding to supplant sight.   And naturally, if the acquisition of such a skill aids the animal in exploiting its environment, giving it a competitive economic advantage over other unskilled animals, it and its progeny will prosper and his skill will be, by definition, adaptive.

Animals, and humans too, for that matter, *know* that they hear. The sensation of sound is passed through the intellect and the sound is analyzed by the mind. We try to deduce what it was which produced the sound, and whether the sound represents a threat or not. But we do not seem to know the rhythm of time. We know whether it is day or not, and we know of vague feelings of discomfort when time is disoriented by transoceanic jet voyages. But we have no precise sense of circadian rhythms analogous to the honeybee's precision, which can predict the sun's azimuth and use this azimuth as part of a communication system.

Once von Frisch had broken the honeybee dance code and discovered that it made use of time as a space beacon, and Lindauer had discovered that this time sense was "endogenous," more work still had to be done. It was not known whether this unique skill of relating space to time had evolved in response to the bee's need for an analog communication system or whether it had evolved in response to a space-orientation need. Also, no one yet knew whether this highly evolved sense of rhythmic time was widespread throughout the animal kingdom.

An Italian zoologist, Floriano Papi of the Zoological Institute of the University of Pisa, had become interested in the sand flea, *Talitrus saltador*. Though called a flea, this tiny creature is not an insect, but a crustacean distantly related to the shrimp. It normally inhabits that wet band of beach sand which is just out of the battering reach of the surf. During the bright part of the day, the late morning and early afternoon, it remains hidden in underground tunnels, obtaining its food by filtering plankton from the seawater which soaks through and collects in its tunnels. But around sunset the sand fleas emerge from underground and embark upon a strange migration, traveling inland for a distance of one hundred meters or more. Papi was fascinated by several aspects of this migration. It appeared to be purposeless, at least in any economic sense. It was not a food search, for the flea is unequipped to ingest any solid food. Differing numbers of animals make the trip each night. During certain periods large numbers of animals make the voyage, while at other times only a few may be found. Papi could not discover any associated phenomenon that correlated with this change in numbers. It did not seem to occur in connection with any easily observed astronomical occurrences, and Papi suspected that perhaps weather-induced barometric or humidity changes may have triggered more fleas at one time than another.

Around sunrise the fleas return to the beach and the water's edge. But if they are interrupted at any point during the course of their migration, they will attempt to escape threat by returning to the sea. Papi found that they returned to the sea unerringly, even when he caught them and transported them to a different location where familiar landmarks were absent. Strong offshore winds scattered the sand and continually reshaped it into different patterns, so that even under normal circumstances it would be difficult for them ever to locate themselves by means of a stable set of Euclidean reference points the way we humans do. Suspecting that perhaps sand fleas sensed the presence of the sea by some stimulus such as the sound of the surf, or differences in humidity, or scent, Papi caught some fleas and transported them across the boot of Italy to the Eastern shore, the Tyrrhenian seacoast, where he released them a few meters inland from the surf line. He was astonished to discover that instead of proceeding directly toward the nearby beach, they struck off the way they had come, apparently determined to march overland for more than one hundred kilometers in order to reach their familiar Adriatic beach.

When von Frisch's honeybee work was published, Papi immediately set out to discover whether a similar solar beacon system operated in the migration behavior of his sand fleas. During a period of thirty-seven nights of August and September, 1959, Papi caught 11,429 sand fleas by setting upright two sheets of glass sunk in the sand to form a V about a yard wide at the mouth. Papi writes that he captured them with a view to conducting the rest of his studies in the laboratory, where control conditions could be more easily obtained. But once in the laboratory they refused to migrate, though he diligently attempted to duplicate all the necessary conditions. At last he was forced to haul his equipment back to the beach and continue his studies there.

He placed his fleas on shallow glass trays in the beach and recorded the course of their movement by means of regularly spaced photographs. He was able to demonstrate that during the sunlit portions of their voyage they navigated either by means of a direct view of the sun, or (if he shaded his trays from the sun) by referring to the vibration plane of polarized light. As von Frisch had done, he used sheets of polaroid plastic to disorient the voyage of his fleas.

Their nighttime voyages were a different matter. Lindauer had demonstrated that the honeybee referred to an innate knowledge of the regularity of the sun's changing azimuth. Papi saw no sign of any such factor operating with his sand fleas. On the contrary, he noticed that though on certain nights they may have begun their voyage choosing their direction from the sun's position, once the moon rose, if it became obscured by clouds the fleas stopped their march and dug themselves into the sand wherever they happened to be. Papi set up his glass trays marked with a grid and began taking flash photographs of the nighttime migrations. When he set up his trays on moonlit nights at the farthest inland point of migration, the animals accumulated toward morning on the seaward side of the trays. But if he shaded the trays from direct moonlight, the animals seemed to lose their sense of direction and became randomly dispersed in the tray. He then experimented with redirected moonlight shone onto them with mirrors, and found that he could divert their course of travel. He could also accomplish this with the beam of a flashlight, even when the flashlight was offered in direct competition with the natural light of a full moon. But these diversionary tricks only worked when the moon was fairly full. "During the first and last quarters of the moon," Papi writes, "the angles [of the flea's travel course] assumed with the mirror are very different from those expected. The cause of this phenomenon is not known."

Papi concluded that, as with the honeybee, in the *umwelt* of the sand flea time and spatial direction are locked together in the consciousness of the creature. Papi found no evidence of any directional communication system used by the sand flea—their circadian rhythms had adapted to serve them purely for spatial orientation. But it was by no means a simple system. In fact, Papi was forced to the conclusion that there existed two circadian rhythm systems in operation. The first was clued, like that of the honeybee, to a twenty-four hour solar day. The second system, according to Papi, has a rhythm "whose period is not however of twenty-four hours, but equal to the duration of the lunar day."

All this work—with fiddler crabs, honeybees, sand fleas, and plants—was fascinating, but it seemed only distantly applicable to men, if it was applicable at all. Man is a vertebrate, and none of the creatures so far investigated were.

It is not my intention here to imply a historical precedence by citing work done in this field. Directly as the phenomenon became known, work proceeded, as it does these days in almost every scientific discipline, along a broad front. However, in the evolutionary order of ascending complexity the next most complex creature shown to orient itself according to circadian rhythms was a vertebrate animal, a fish, the common Wisconsin green sunfish. And a most elegant demonstration was performed by two zoologists; Wolfgang Braemer, who had temporarily transferred from the Max Planck Institute at Seewiese and was working at the University of Wisconsin, and a graduate student who worked with him, Horst Schwassman. Braemer was puzzled by the whole idea of solar navigation. He wrote: "The first question which arises in sun-orientation is: what possible information given by the sun is really used by the compass animal? Generally it has been assumed that only the azimuth of the sun's position, that is, the projection of the sun's position onto the horizontal is used. The sun's azimuth however, changes during the day with varying speeds. In order to maintain one compass direction accurately, the correction of the azimuth movement has to vary with the same rhythm. The azimuth movement varies not only during the course of the day, but also with the seasons, and in addition, they are different at different latitudes. An animal which uses only the sun's azimuth for keeping a compass direction has therefore to compensate different azimuth movements in equal time intervals at different times of day at different seasons and at different latitudes if it migrates."

Braemer and Schwassman then built a large, shallow, circular tank out from the shore of a lake in Madison, Wisconsin. In the center of this tank, they installed a thick tray like a Lazy Susan, capable of being rotated so that no surface imperfection such as a scratch or a blister of paint could serve as a landmark clue to the fish. Fish placed in this barren tank were most uncomfortable; there were no places to hide, no protective shadows, no crannies. Knowing that fish need a place to hide in the outdoors, Braemer accommodated them; that was the whole point of the experiment. At one point on the circumference of the Lazy Susan there was a shelter box, but the fish would have to swim out to the very edge of the Lazy Susan in order to find it.

What Braemer and Schwassman did was to release a fish in the center of the tank and watch it swim hectically around the tank trying to find shelter. Sooner or later it would discover the box and hover inside. Once the location, that is the compass direction of this box in relation to the center of the tank, was learned by the fish, it would dart straight off in that direction as soon as it was released in the center of the tank. Both the Lazy Susan and the tank itself were rotated in relation to the shelter box in order that no landmark clues could be offered to the fish—even Braemer and Schwassman concealed themselves, observing the activity through a system of periscopes. They noted that the fish seemed confused in the absence of the sun on cloudy days, but this was not yet proof positive that the sun was being used as an orientation beacon. After their fish had been trained outdoors in natural sunlight, Braemer and Schwassman moved their tank into a dark shed. They posted a bright light off to one side of the tank, and then at various times of day released trained fish in the center. Naturally, the fish did not find the box on their first attempt, but by plotting backwards from the course of their initial search to the position of the sun at the time the test took place, and the relative position of the electric bulb, Braemer and Schwassman were able to demonstrate with convincing elegance that it was indeed the sun's azimuth which was providing the fish with orientation clues.

As the last ornament of their demonstration, in an attempt to prove that solar orientation is innate and not learned, that sunlight neither acts as a *zeitgeber* nor as a teaching tool, Braemer and Schwassman raised twelve sunfish from the egg under continuous light. When they were mature, they were trained to locate their shelter box while the tank was in the shed by means of an electric light placed approximately where the sun would be if they had been outdoors at the time of their training. They were given thirty trials in a fifteen-minute period. Several hours later the tank and the fish were brought outdoors, and the fish were released to find their way to the shelter box by their inbuilt expectation of where a sun they'd never seen should be in a sky they'd also never seen. Of the twelve fish, nine swam immediately in the general direction of the shelter box, two fish were completely disoriented, and one attempted to use the sun as a fixed point of light.

Several years before Braemer and Schwassman performed this elegant and convincing demonstration, a zoologist named Gustav Kramer had stumbled over something curious. In 1949 Kramer kept a group of starlings in an outdoor aviary. He was interested in avian sociology—how bird flocks formed themselves, how certain birds obtained leadership roles and dominant status—and so that there would be ample space for individuals to move about, his aviary was quite large. He began to notice, as summer turned gradually into fall, that his birds seemed to be becoming more restless than usual.

During the evening they congregated off to one side of the aviary and began to take short fluttering flights, banging into the wire mesh and returning to their perches giving every indication of extreme unhappiness. Kramer recognized a "migratory restlessness aimed in the direction normal for migration of this population." Immediately he dropped his interest in avian sociology and proceeded to consider the possibilities now opened by this simple observation. As one of his associates later wrote of it, "this discovery of directed migration in a restricted space allowed him to overcome the main difficulty that had hitherto hampered experimental work on long distance orientation, namely the long distance itself."

Kramer began by designing a circular cage within which was fitted a circular perch. This perch was marked off in compass point degrees and careful records were kept of the locations on this perch where the bird seated itself, and the duration of its perching time. After a considerable number of observations, when these were plotted on a graph, it became obvious that even though caged, the birds showed a decided directional preference. Moreover, this preference appeared to coincide with the general compass point direction of their habitual migration.

Unfortunately, Kramer died before he could fully exploit the ramifications of his discovery, and his work was continued by a husband and wife team of colleagues, Franz and Eleonore Sauer. They made some minor modifications in the original Kramer cage: they reduced its dimensions, so that it now measured only about a yard in diameter, constructed it out of plexiglass instead of wire mesh, and provided false inner walls of yielding, nearly invisible nylon net so the birds would not injure themselves banging against the enclosure during abortive flights.

Then they chose a likely bird for their experimental subject. It had to be of moderate size, it had to migrate over a large distance, and preferably should habitually make this migratory trip alone, rather than in a flock. They chose the lesser white-throat, a common garden warbler which summers in northern Germany and winters in the southern Congo and northern Rhodesia area of Africa. It migrates alone and at night, and though weighing barely three-quarters of an ounce, it travels with great speed, averaging over one hundred miles a night—flying a fairly true course, banded birds being found only along a rather narrow flyway.

The Sauers began their work in 1958, capturing summer nest-lings and raising them by hand till the autumn. Then they exposed their hand-reared birds in their plexiglass cages to the night sky over Bremen. The transparent walls of the cage were covered by black felt so the birds could see no terrestrial land-marks.

They found that the birds showed a preference for south-southeast, the direction of their migration route, only when the night was clear enough for them to see the stars. On cloudy nights when only the moon was visible, or when bright city haze destroyed the *gestalt* pattern of the sky, the birds showed a random pattern of activity.

Their suspicions that the stars were acting as navigational guides thus reinforced, the Sauers then proceeded to the next step of their experiment. They found a small Zeiss planetarium operated in Wilhelmshaven by the German merchant marine to teach the rudiments of celestial navigation. They brought their birds, inside their cages, into this planetarium. The first night the birds were shown the simulated sky over western Germany. The following night the planetarium projector was adjusted to show a star picture appropriate to a location about one hundred miles south-southeast, and so it went.

One night, as the Sauers began the planetarium session with the simulated Berlin sky overhead, and the birds headed south-southeast, the Sauers switched the projector backwards five hours and ten minutes, thus presenting the birds with the sky "as it would correspond to a point near Lake Balkash in Siberia." The lesser whitethroat thereupon took up a position on its perch pointing due west, a course which would bring it back to its starting point in western Germany. The Sauers gradually reduced the displacement of the apparent sky till it "corresponded to a position near Vienna," at which time the bird assumed a new directional perch, facing in a south-southeasterly direction.

Interested to determine, if they could, by which particular stars the birds navigated, the Sauers turned off various lights within the projector, thus experimenting with slightly distorted skies. They found that it was the general pattern of the brighter stars (not the planets) that guided the birds, and great numbers of lights could be omitted without unduly disturbing them.

As plotted in the planetarium, the birds chose a route which would if traveled across our planet, take them over Italy, down the length of its boot, and out over the Mediterranean to the northern coast of Libya. Their course was that of the great circle route— plotted on a Mercator projection map, it would take the shape of a bow, bulging slightly toward the east. The birds would fly over the western Sudan and then turn ever so gently south-southwest to land eventually in the lower Congo.

One of the residual puzzles left unanswered by the Sauer experiments has to do with the fact that the real flyway of the birds in the wild, as plotted by trapping banded birds, is quite different from the flyway as plotted in the planetarium. Apparently the wild birds are deflected by the Alps: no lesser whitethroat has ever been seen in Italy. It is also doubtful that the birds fly over the Mediterranean; they turn eastward instead, flying over Yugoslavia, Bulgaria, Turkey, and then turn abruptly southwest to fly over Egypt, into the Sudan and the Congo.

Critics of the Sauers attacked their work on the basis that perhaps the earth's magnetic field exerted an influence on the birds, and their reponse to the planetarium starshow was only apparent. The experiments were repeated the following year, this time using two planetaria, each in a different location in

Bremen, not Wilhelmshaven. Each planetarium had a magnetic field different from the other but the results were identical. In addition to the lesser whitethroat, the Sauers also used other species of birds—the lesser grey shrike, the wood warbler, and the blackcap. Each of these night-migrating species used the stars for navigation. In the planetarium plots of each species there was some inexplicable variation between the planetarium route and the actual route taken by the birds in the wild.

This suspicion that animals respond to the directional pull of the earth's magnetic field has nagged zoologists for years. There have been many inconclusive experiments directed at this inquiry. One example was conducted by Frank Brown who decided to test a bit of New England folklore which claimed that the fast-creeping New England mud snail, *Nassarius obsoleta* will respond to a magnet.

Brown captured some mud snails and devised a contrivance to test their magnetic sense. He built a tray in the shape of a fan or a slice of pie with an acute angle of about forty-five degrees at the point. The snail was released at the point of the fan. No stimulus was offered; the snail was allowed to wander randomly on the tray. The snails found the surface of the tray and their exposure on it uncongenial, and usually chose a fairly straight path, hoping to get off. Brown shifted the tray in various directions relative to the earth's magnetic field, plotted the tracks made by the snails to see whether there was any statistical preference in course depending on the orientation of the tray. He also created artificial fields by placing a heavy alnico bar magnet under the tray and repeating the experiments. He found, as he had hoped he would, that there was indeed a distinct change in the courses chosen by the snails to get off the tray. But to his amazement he found that the snail's response to magnetic fields was not constant. It varied considerably and *predictably* during the course of the day in relation to the position of the sun! Brown writes: ". . . the two orientations of the experimental magnetic fields gradually change their effectiveness as a clear function of the sun's angle. Described in general terms, the parallel [to the earth's] field is more effective in turning snails when the sun is high in the sky, the right angle [to the earth's] field more so at night, with the two fields about equal near sunrise and sunset. The magnetic receiving system behaves," he continues, "quite as if it comprised two sets of 'directional antennae,' one geared to the solar day, and the other to the lunar day." What role this magnetic sensitivity plays in the purposeful life of the mud snail, no one knows. When Brown began working, he doubtless hoped that the snail's perception of the earth's magnetic field would prove to be an orientation system, used in much the same way as humans use a compass. But a compass is only useful if it is constant. North must be north if the time be high noon, or midnight. But this may be merely our clumsy way of understanding sensory deficiencies which we usually think of as being constant limitations. Though our human hearing is limited to a comparatively narrow band of wavelengths, so far as we know, this is constantly so. It would be strange indeed if at certain periods our perception of marginal sensations (such as the magnetic field perception of the mud snail) would be expanded, only to contract again at other periods.

Yet strange as this may seem to us, it was not always considered an impossibility. All the myths of mankind are replete with suggestions that there are certain holy days which cluster (as one compares one religious system with another) around certain points of the calendar. And the ceremonies which are performed on those appointed days invariably involve the hypothesis that on those days the communicant in the ceremony is especially responsive to those "cosmic" energy forces which bind the universe together as a coherent system. Traditional anthropological conceptions consider these ceremonies to be responsive; that is the ceremonies are understood to have originated as celebrations, grateful responses to the fertility of the land, or the breeding cycle of game animals, and so on. That there is a large element of thanksgiving in these ceremonies is without doubt entirely true. But there must be more. The basic rhythms of day and night and the passage of the seasons affect all living cells in much the same way. Despite the evolution of man's mind, the brain itself is still a mass of living cells. As such it, too, is responsive to periodic events.

Our technological control of our environment puts us into an ambiguous relationship with that environment; similar to that odd and destructive relationship that arises between a jailer and his prisoner. It seems as though the more control the jailer exercises, the more a strange kind of perverse love, a love that thrives on injury, grows between him and his charge. As the relationship develops, the prisoner often becomes the stronger of the two, inflicting by his very passivity the greater hurt.

So we, as we control the environment, have gradually become the victims of our own control. The role of cycles in our lives, being natural, should be a joyous source of strength. Generally, however, we do not acknowledge their existence at all, and when we do, we see these cyclical changes in ourselves as impediments to our efficiency.

For example: most plants and animals inhabiting the temperate zones tend to inhibit their activity during the long nights and bleak days of winter. Yet we human beings have—just recently, and before our bodies have been allowed that expanse of time for evolutionary adaptations—made this period the busiest of the year. We have come to organize the apex of our professional and social busyness around the darkest point of the annual cycle, that time immediately before and after the winter solstice which occurs on December 22nd. No wonder then, that the period immediately following this is the favored time for an increased incidence of disease and a rise in the mortality rate.

Another obvious example of our disregard of that which everyone knows and feels: we know that flying from one hemisphere to another across time zones is disruptive to biological rhythms. The inner time sense of the individual must reorganize itself in response to external time cues. Under natural conditions these changes are sufficiently gradual as to make the transition easy. However when six or more hours are suddenly lost or gained, the system is greatly disorganized. There are some physiological effects, but by far the greatest number of disruptions manifest themselves in feeling, mood, and psychological functioning. Everyone who has jetted across hemispheres knows that upon arriving at one's destination, the greatest need of the body and mind is for rest and tranquillity. Yet so thorough is our rejection of these needs, that it is precisely at this time that it is customary to schedule the most hectic and exhaustive round of activities. And this willful avoidance of considering biological or cyclical time becomes particularly pernicious in business or diplomatic confrontations, where one party has traveled a considerable distance. Both parties arrive at the conference table pretending a total equality. Yet the resident host has a tremendous psychological advantage over the guests who arrive at the table distracted and easily fatigued. They can neither concentrate with their accustomed tenacity, nor do they have their customary stamina.

Partly, this disregard of this literally painful biological fact of life stems from the ancient Judeo-Christian dichotomy, the separation of body and mind into quite separate compartments. The bodily effects of such traveling is well known to athletes who usually schedule their arrival well in advance of competition so as to allow a period of recuperation. Yet businessmen or diplomats often allow themselves no more than a quick sprucing up at a hotel before being whisked off to meet the opposition.

The mind is as much in the body as the body is in the world. The body penetrates the mind just as the world penetrates the body. We like to believe, since we see ourselves as enclosed within a shield of skin, that we are demarcated from the world by this envelope of skin, just as a theater curtain separates the audience from the stage before the performance. But the skin is a porous membrane. Electrically and chemically the world moves right through us as though we were made of mist.

By and large we are unaware of the presence of the outside world within us. We are even more unconscious of the breathing of the skin's pores than we are of the intake and exhaling of the lung's breathing. We do not feel the penetration of cosmic particles. This part of the world is all but unknown to most of us. And yet it is as the world enters into *us*, with its force and influence, that we become one with *it*.

Some exceptional people occasionally have this sense of the "seamlessness" of the unity of the world. They are known in the West as mystics. Others have it all the time, and they are known as schizophrenics, a portmanteau description of psychic states which render their possessor relatively unfit to deal with the "real" world of matter, routine, and interpersonal activities. Many of us have certain mystical experiences which occur in dreams and perhaps in fantasies at certain cyclical periods. In other civilizations, and in other historical times in our own civilization, when these psychic states were judged to be less pathological than they are now, they were induced by drugs, fasting, or strenuous ritualized activities. In some religious systems they were induced by contemplation practices, which perhaps produced the effect of sensory deprivation, which we now know through laboratory tests, can drive a man "mad."

Yet it was the closed-system logic of the West which produced its technology. Each system was examined as though it were closed, and then the fissures in its closure further examined until the essential mysteries became ever more microcosmic or macrocosmically removed from the commonplaces of life. For the average electrician the ultimate source of power is the electric utility company dynamo; he is not continually confronted with the unknowable nature of energy.

The same closed-system logic which produced technological achievements operated in medicine as well; and in the beginning medicine progressed by resorting to a dualistic separation of body and mind. It was René Descartes who, in the seventeenth century, laid the groundwork for all these advances. He stipulated that there was a distinction between thought and matter, whether that matter was external to the body or internal within the body itself. But even he, emphasizing as he did the duality of body and mind, was too astute not to hypothesize a connection somewhere. He searched for a kind of physiological valve, or gate, through which the perception of matter might be transformed into thought. Lacking the techniques and tools for such research, he relied upon the literature and settled on an odd little mass of cells—located at the base of the brain and called the pineal body—as the physiological site of what he called "the rational soul."

The pineal body is a small grey or white structure about a quarter of an inch long and weighing about a tenth of a gram in man. It is located at the very top of the spinal column where the neck enters the skull. It is the only structure in the brain which is not bilaterally symmetrical. Draw a mid-line down the brain from front to back and everything appearing on one side of this mid-line is duplicated on the other—except for the pineal body. This fact alone has always made it distinctive and a curiosity to anatomists. It is shaped roughly like a pine cone—therefore its name. Descartes was not entirely original in his description of the pineal body as the site of the soul. The Greek anatomist Herophilus of the fourth century B.C. described the pineal as a "sphincter which regulated the flow of thought."

Herophilus in turn may very well have come by his notion of pineal function from India, where speculation about this apparently useless appendage to the brain stretches back into the darkness of prehistory, perhaps as far as 3,500 years.

The Hindus recognized something about the pineal body which had escaped the intuition of Western anatomists from Herophilus to Descartes right up to the year 1886, when by apparent coincidence two monographs on the subject were published independently, one in German by H. W. de Graaf, and one in English by E. Baldwin Spencer. The Hindus recognized from the very beginning that the pineal body was an eye! As such it is represented in oriental art and literature as the Third Eye of Enlightenment.

The third eye of certain lizards and fish was well known; the eye was unmistakable in the *Sphenodon* genus of lizard, which contains the famous species *Tuatara* of New Zealand. In this creature the third eye is marked distinctly by the skull being opened in a central cleft, and the external scales arranged in a kind of rosette with a transparent membrane in the center. When dissected, this organ proved to have all the essential features of an eye—there was a pigmented retina, which surrounded an inner chamber filled with a globular mass analogous to a lens, but the connecting nerves were absent, and the anatomists of the nineteenth century decided that the organ was without visual function in the *Tuatara*. It was primarily De Graaf and Spencer who proved that this organ in the *Tuatara* lizards was the same organ that became buried and was rendered indistinct in function in mammals and was known as the pineal body.

In 1958 two zoologists, Robert Stebbins and Richard Eakin from the University of California at Berkeley, did ethological studies on the common western fence lizard (*Sceloporus occidentalis*), capturing two hundred animals, removing the pineal eyes of one hundred, and performing a sham operation equally traumatic, but which left the pineal eye intact in the other hundred. They found that removing the pineal eye markedly affected behavior, particularly escape reactions. After recovering from the operation, the animals were released in their natural habitat, whereupon Stebbins and Eakin chased and tried to capture them. Before 10 A.M., they were able to capture 63 per cent of the pinealectomized lizards as opposed to 37 per cent of the sham operated animals. After 10 A.M. and until nightfall the percentage of captures was more equal, though a small balance of the sham operated animals always managed to escape better than the pinealectomized ones. This experiment proved that, contrary to all the previously held notions about the vestigial nature of this third eye—that it was a remnant as useless to lizards as the appendix supposedly is to humans—possession of an intact pineal eye has a marked survival value. It was most emphatically *not* nonfunctional.[1]

In mammals the pineal body was recognized to be a gland by the early Latin anatomists, who rated it more important than the pituitary, which is now considered the "master gland." The Latin writers named the pineal "glandula superior" and the pituitary "glandular inferior," but from the nineteenth century until just within the past ten years modern anatomists have been divided into two distinct camps—those who believed the pineal was an endocrine gland producing some hormonal excretion, and those who believed it was a useless vestigial appendage of the brain, a leftover reptilian eye. It was not until 1958 that the secretion of the pineal gland was isolated and identified, and the importance of the pineal as a time-sensor in humans finally established. Strangely enough this tremendously important work was done by a dermatologist, Aaron B. Lerner of the Yale University Medical School, and not by an endocrinologist. The story of how this came to pass is sufficiently curious to warrant a brief diversion.

After Lerner's work the pineal can properly and legitimately be called a gland, not merely a body. But its role as a gland is probably a comparatively recent evolutionary metamorphosis. In many fish, for example, the skull is opened, as with the *Tuatara*, and covered at that spot by a special area of transluscency. The gland lies right underneath the skull, atop the brain. In some birds, and in rabbits and mice, the gland is also located at the top of the skull. In humans however, it has migrated underneath the brain, and it would appear much as it does in Hindu art on a level between the "real" eyes, only it is, of course, buried in the

---

[1] There was no suggestion implied in this experiment that the pineal eye was an image producer. Stebbins and Eakin concluded that it acted as a kind of solar thermostat which controlled metabolism—that the pinealectomized animals were sluggish with cold in the early morning.

center of the skull. In rabbits and mice the root of the pineal is located roughly analogously to that of humans, but the gland itself is stretched upwards by means of a long stalk. Size varies. Elephants and whales have minute pineal glands. Walrus have huge ones, almost an inch long and half an inch wide.

However it appears in size or location, it is one of the most ancient parts of the vertebrate brain, and among existing creatures it is most easily studied in the living forms which have not changed significantly from their ancient forebears. One of these is the *ammocoete*, the larval form of the lamprey. Lampreys are unpleasant creatures; eel-like in shape, the adults are parasites. Their technical name *cyclostoma* derives from two Greek words, *kuklos* meaning round, and *stomax* meaning mouth. That is their most distinguishing feature; their mouth is a huge round sucker, armed inside with rows of needle-sharp teeth. They attach themselves to a fish, and by grinding their mouths back and forth through a small arc, they bore through skin and scale to the flesh, where they suck their host dry of vital fluids. They are primarily fresh-water creatures, but some species spend part of their life in the sea. Invariably, though, like eels they return to fresh water for spawning. They gather in gravel-bottomed, fast-moving water, males and females both, and both pick up stones with their sucker mouths to make a nest. As many as thirty or so individuals may cooperate in this labor. The females then lay a large number of eggs, up to several hundred thousand, which are fertilized by the males, and in a short time, perhaps ten days, the larvae hatch. Known as *ammocoetes* (from *ammos* meaning sand and *koite* meaning bed in Greek), they so little resemble their adult form that for many years they were thought to be different animals entirely. During the extensive metamorphosis into adulthood their heads are partially destroyed—skull, muscles, nerves, pharynx—as is their intestinal tract, and then remodeled. They live as larvae for up to three or four years, buried in the sand as sedentary as clams, filtering sea water through a mass of cilia. In this state they are eyeless, that is they lack conventional

eyes, but they are equipped with a large and apparently functioning pineal eye. In fact they have two pineal eyes, one lying
underneath the other, both beneath the same point in the skull,
the lower more degenerate in form than the upper, but both
joined by separate stalks to the same root. The larval lamprey
changes color in response to external light conditions, becoming
sandy pale in bright light on a sandy bottom, and altering to a
dark muddy brown when conditions for this color change are
appropriate. Even before the turn of the century this facility
was noted in ammocoetes, and since they lacked conventional
eyes, it required no great leap of the imagination to construe that
the pineal gland served both as a light sensor and as an organ for
altering skin color.

Several other reptiles change color in response to light conditions, in addition to the well-known chameleon lizards. Many
species of frogs and toads, especially in their tadpole stage, turn
a deeper color in darkness, though their change in color is not so
spectacular as that of the lizard and requires a careful observer to
note it at all. But with frogs, skin color is also affected by
emotional states. A frightened frog is a light-colored frog. It
was this factor which drew the attention of Dr. Lerner to frogs
and tadpoles. As a dermatologist he was interested in human
skin color—how and why it changed in shade in such diseases as
vitiligo, or piebald skin (an affliction marked by the appearance
of white scabrous patches), and in melanomas (dark-colored
growths akin to warts), which may often be malignant.

Around the end of the nineteenth century and the beginning of
the twentieth, a great deal of fruitless work was done with the
pineal gland, when it was first identified as being a vestigial eye.
Experiments were conducted on all manner of reptiles, and in
1917 two American zoologists, Carey P. McCord and Francis P.
Allen, began working with tadpoles, feeding them with pineal
gland extracts and watching them turn light. The strange and
intriguing fact about McCord and Allen's work was that they
discovered that this pineal substance—whatever it was—crossed

species barriers. They could feed frog pineal extract to their tadpoles and watch them turn light, and they could feed cattle pineal extract and watch them become equally pale. Over the forty intervening years between McCord's work and Lerner's, this color response to pineal extract had been repeated untold times in classrooms and laboratories, but no one had even come close to isolating or identifying the substance which worked this change.

Lerner began by making arrangements for the Armour and Company slaughterhouses to supply him with the pineal glands of cattle. During the course of the next four years, which he spent on this project, he used over 250,000 glands. No records were kept of the numbers of tadpoles and frogs, but they must have numbered nearly as many. In 1958, after incredible labor and the most ingenious use of equipment and techniques, Lerner and his associates finally isolated and identified a new hormone manufactured only in the pineal gland. One of its properties was that it inhibited the spread of melanin in the melanocute or pigment cell. This property of the substance suggested to Lerner an appropriate name, *melatonin*, a coinage from the Greek meaning literally "darkness constricting."

With this work the 3,500-year argument over whether or not the pineal was a gland was finally settled for good. It was a gland, there was no more doubt. And with this discovery investigators surged ahead to try and find out why melatonin existed in the body, and what it did. After all, cattle do not change color in response to light; melatonin must have another function in the body economy of mammals.

Lerner's labors had a sadly ironical ending for him personally. His discovery did him no good at all in answering those questions which he had wished to pose to nature. Melatonin has seemingly nothing to do with changes in human skin color; he is no wiser now about the possible causes of skin cancer than he was when he began. Investigators who are using his remarkable discovery as the jump-off point for their own work send him reprints of their publications. He reads them rather wistfully in his office, where he recently told a visitor, "It's nice to be able to contribute to another man's field, but it would be even nicer if I could feel I'd done something worthwhile in my own."

According to traditional medical doctrines, glands were considered to be dependent upon the bloodstream, upon substances carried in the bloodstream which triggered the glands to produce other substances for the body's needs. A corollary to this theory held that glandular hormones were carried in the blood in generally constant quantities. When these quantities were decreased by a sudden demand for them on the part of tissues that required them, the gland somehow sensed the lack of these hormones and quickly produced more of them.

Science has a convenient way of packing its theoretical suitcases. It just leaves odd bits of things hanging out—the foot of a sock here, a show of shirttail there—but so long as the suitcase closes and can be carried about, it serves its functional purpose. And after a while, no one seems to notice these tag ends of things that won't pack.

But this particular theoretical suitcase would never even stay closed, for there was one eminently normal, regularly occurring situation that didn't fit the theory, and that was the periodic menstruation of women. At this time in women all manner of dramatic endocrine changes become apparent. There is nothing subtle about them; they are as spectacular as fireworks. They couldn't possibly be ignored. And yet they were, for whence did the stimulus come? The sudden onset of fear or horror, for example, can produce endocrine reactions; the whole body clamors for internal as well as external rescue. But what is the stimulus which produces the endocrine changes accompanying menstruation? Time itself? If there were some external time stimulus in the environment, the phase of the moon, or the tidal patterns in any given locality, then theoretically all women within the range of this phenomenon would react simultaneously. But such is not the case. Though each normal woman's period is constant, from woman to woman the period is totally random. In the human the female alone responds sexually to cyclical time. But in the lower forms males and females both, in many cases, are sexually cyclical in both physiology and behavior.

Botanists had long recognized a clear connection between light cues and sexual behavior (flowering) in plants. By artificially altering the hours of light to which plants were exposed, premature sexual development could be induced. In animals the same phenomenon had been observed. Animals which came into rut in the fall could be induced to believe that the fall season had come round even when it hadn't, if their artificial light was turned on progressively later in the morning, and off progressively earlier in evening.

In 1898 there was published by a German physician, Otto Heubner, the first report establishing a connection between human sexuality and the pineal gland. Heubner discussed in his report the case of a young boy with precocious puberty—overdeveloped gonads, pubic hair, and so on—who was discovered to have a tumor of the pineal gland. Since then this peculiar association of symptoms has been reported a great many times but almost always in connection with young boys, not girls.

After Lerner's discovery of melatonin, and his finding that though it had an effect on reptilian skin color, it seemingly had no known effect on mammalian systems (including skin color), there was a concerted effort to discover the use that the body made of melatonin. It was well known in 1961 that if rats were kept in continuous light, the estrus cycle of females was speeded up and also the size and weight of their ovaries was increased. In that year Virginia Fiske, working at Wellesley College, exposed rats to continuous light, killed them, weighed their pineal glands, and found that pineal gland weight decreased in comparison with a control group kept in alternating light and dark conditions. Here were Heubner's findings with human males confirmed in female rats. In Heubner's boy one would suppose that the pineal tumor caused glandular secretions to cease. The result was premature sexual development. The secretion of the pineal gland, therefore, could be considered to inhibit sexual development. In Fiske's rats increased ovary size was accompanied by a decrease in the weight of the pineal gland, and it could legitimately be assumed that the sexually inhibiting secretions of the pineal gland were retarded.

This was the first clue to a possible role, in the body economy of mammals, for Lerner's substance. The next step—which was being taken at this very moment by two biochemists, Julius Axelrod and Herbert Weissbach, working at the National Institutes of Health in Bethesda, Maryland—was to discover how the pineal manufactured its melatonin. Axelrod and Weissbach succeeded in recreating the chemical transformation and found the enzyme responsible, but in so doing, opened the door into a new and even more puzzling corridor of mysteries.

Axelrod and Weissbach had discovered that the enzyme which they named HIOMT worked upon the molecules of a precursor chemical, serotonin, in order to produce melatonin.

Serotonin is shrouded in mystery. Even the history of its discovery is mysterious. It would seem that an Italian pharmacologist, V. Ersparmer of the University of Rome, published in 1946 the first report of a substance which he had found in the intestinal walls of vertebrates and in the salivary glands of octopi. It seemingly had a property that caused contractions in smooth muscles, and believing that it had something to do with the peristaltic movements of the gut, Ersparmer named it enteramine. He published a report of his work in an Italian journal of biology where it was largely unread by English-speaking scientists.

Two years later Maurice Rapport, a hemotologist working at the Cleveland Clinic, while searching for that substance in blood that promotes clotting and causes constriction of veins, isolated and produced an artificial crystalline form of a substance, which he called serotonin. It tended to cause blood to clot, and to promote the constriction of smooth muscles. Eventually biochemists found out that it was identical to Ersparmer's enteramine.

The strange story of enteramine-serotonin only begins at this point. It was soon discovered to be a commonplace chemical in nature; Ersparmer had found it in the salivary glands of octopi. It was also found in many plants and vegetables. Bananas and plums abound in serotonin; so do figs, and among the species of figs none is richer in serotonin than the *ficus religiosa*, known in India as the Bo tree, under which the Buddha reportedly sat when he became enlightened.

Though halfway round the world from this moment and some 2,500 years removed in time, there was a grossly analogous occurrence in Basle, Switzerland, in 1948. In that year a dour pedantic Swiss chemist, working in the laboratories of the Sandoz Pharmaceutical Company, bicycled home for lunch and did not return that afternoon. His name was Albert Hofmann; he had been working with chemical compounds derived from ergot, a fungus found in rye plants. He had somehow absorbed a minute amount of this substance into his bloodstream, and he did not leave his home that afternoon because he thought he was going insane. He had discovered one of the effects of the compound he was working with, lysergic acid diethylamide, later abbreviated to LSD.

In the ensuing twenty-odd years this chemical, now abbreviated to LSD-25, has become a psychiatric and pharmacological puzzle of major dimensions. That mental states could be altered by ingesting various drugs is nothing new to mankind. But the effects of this drug were so profound from such minute amounts, that it demanded investigation. In some subjects the drug appeared to imitate the symptoms of mental disease, paranoia or schizophrenia. In others, religious experiences analogous to those reported in the mystical literature of both East and West were a commonplace. They include, in addition to the sense of brilliant light, a feeling of euphoria, of oceanic peace and happiness, true blissfulness, and a sense of oneness, of being atoned with the universe at large. All sights and sounds, all sensations external to the self, become incorporated into the self. The sense of alienation which is so much a part of human existence was annihilated. As reported by some who underwent the experience, there was a strange sensation that the self flowed out into all other things, and conversely all other things flowed into the self. For some subjects this feeling of total identification with the world brought an ecstatic rapture, while by others it was perceived as a loss of personal identity and became terrifyingly threatening. Some subjects were reduced to primordial horror and panic. They huddled into a corner shrieking.

The molecule of LSD-25 is remarkably similar in structure to that of serotonin, and as soon as Rapport had announced his discovery of this latter substance, the search was on in all major medical laboratories for antimetabolites, which would block serotonin's actions. There are many medical situations in which clotting of the blood and constriction of the veins are undesirable, and it would be useful to have a compound in the physician's armory to counteract these effects.

It was a professor of Pharmacology, John Henry Gaddum of the University of Edinburgh, who seems to have been the first to note that LSD-25 was a potent antagonist of serotonin. His findings were confirmed and extended by the great blind biochemist of the Rockefeller Institute David Wooley, who wrote in 1954 that he had tested various substances antagonistic to serotonin and that "among each of these classes of compounds are some that cause mental aberrations. . . . If this be true, then the naturally occurring mental disorders—for example schizophrenia —which are mimicked by these drugs may be pictured as being the result of a cerebral serotonin deficiency arising from a metabolic failure rather than from drug action." Wooley was saying, in other words, that it was not the direct disruptive action of the LSD molecule itself which caused aberration; it was an indirect action. The LSD molecule produced its effects by depriving the brain of its serotonin content.

At this point the medical profession went through a spasm of grandiose hope. It seemed obvious from these experiments that there was a chemical basis for mental aberration, and that one might hypothetically provide a "cure" for schizophrenia if only one could replace the brain's depleted supply of serotonin. The same thought had struck Wooley, of course, and he reported that he was disappointed with his first attempts. "Experiments with animals revealed," he wrote, "that serotonin injected peripherally fails to penetrate the blood-brain barrier." Many subsequent attempts, using all the various means at the pharmacologist's disposal, have also failed.

Since serotonin did not enter the brain from the body, there must be a local site of serotonin manufacture in the brain itself. Axelrod and Weissbach had proved that melatonin, the pineal hormone, was manufactured in the pineal by enzymatic alteration of the serotonin molecule. Serotonin must then perforce exist in the pineal. Was there another site of serotonin manufacture in the brain, or was it, like melatonin, manufactured in the pineal itself?

Investigations designed to provide answers to these questions were undertaken by two professors at the Yale Medical School. Daniel Freedman, a psychiatrist, had been entranced all his life by the Cartesian body-mind dilemma. Here was a biochemical problem situated in the very organ that Descartes had pointed to as the "valve" between the body and the soul. Occupying an office across the hall from Freedman's, was a pharmacologist, Nicholas Giarman, who had spent a year at Edinburgh as a student of Gaddum, the man who first noted the connection between LSD and serotonin. The men began working together.

They made arrangements with the staff pathologist at the Fairfield Hills Hospital in Newtown, Connecticut, to have a resident physician, Luis Picard-Ami, who was trained by Giarman, perform serotonin bio-assay tests on the brain tissues of deceased mental patients. The team imposed cautious restrictions on themselves. They performed their tests on fresh cadavers, but since permission to perform autopsies is only obtained after some delay, in a nine-month span they were only able to examine the brains of eleven subjects, eight of them mental hospital patients, and three automobile accident victims who had no known history of mental disease.   They discovered immediately that the pineal seemed to be the serotonin reservoir for the brain as a whole.   Measured in micrograms per gram of tissue, in the pineal there was an average of 3,140 units as opposed to the next richest site of serotonin in the brain, the gray section of the midbrain which contained, on the average, only 482 units.   There was another surprise in store for them as they proceeded.   They found a tremendous jump in the pineal levels of serotonin in primates as opposed to other mammalian forms.   "In fact," they wrote, "the serotonin levels found in the majority of human and simian glands are the highest ever reported for any neural structure of any species examined."

And this still was not all.   Bovine pineals are fairly constant in their serotonin levels, ranging in those examined from .20 micrograms to .63 micrograms.   There was a leap in serotonin quantity variation in rhesus monkey glands which ranged from 1.28 micrograms to 3.28 micrograms.   This also represented an exceptional range of variation but it was nothing compared to what Giarmin and Freedman discovered in their human subjects. One thirty-seven-year-old male diagnosed as a chronic schizo-phrenic had a total of only .59 micrograms in his gland, while a forty-five-year-old male sufferer from delirium tremens possessed a total of 22.82 micrograms in his gland.   This, the largest amount discovered in any of the eleven autopsies, was 390 times greater than the smallest.   In terms of gross body weight it was as though from among eleven subjects one giant weighing over 31,200 pounds were discovered along with a jockey-sized man weighing only eighty pounds.   Individual variation between separate members of a population is to be expected, but it is rare indeed in nature to find variations of such magnitude.

However, one must also keep in mind the minute quantities of matter being measured in assaying these substances.   Most of us are familiar with the weight elements associated with an ounce

—an airmail letter, a pinch of salt, etc. But there are more than twenty-eight grams in an ounce. If an airmail letter weighing one ounce were cut into strips, one gram would be represented by a strip about the size of a cigarette paper. And we are dealing here with micrograms, one millionth of a gram! The dose of LSD that totally altered the consciousness of Albert Hofmann as he inadvertently inhaled it was probably around fifty micrograms —fifty millionths of a gram. This is considered a moderate dose in clinical use. Laboratory assay procedures become highly critical when one deals with such infinitesimal quantities of stuff.

Hard clinical correlations between states of mind or emotions and the serotonin content of the brain are yet to come. Actual measurement of the serotonin content of brain tissue can be presently made only at autopsy. As of this writing, there are no really adequate procedures for measuring brain serotonin either via the blood or the urine. As it is metabolized and becomes part of the body wastes, its chemical remnants can be measured in the urine. But what percentage of this total metabolized serotonin came from the brain? After all, since the major portion of body serotonin lies in the gut walls, a rise in urine serotonin metabolites might represent merely an upset stomach, and not the onset of a schizophrenic fugue.

The manner in which LSD alters states of consciousness is also not well understood. LSD is described as an "antagonist" of serotonin, but it does not exert this antagonism in a test tube— only in body tissue. It seems now—as the result of electron microscope studies being made by Freedman and his associates with radioactivity tagged LSD molecules—that LSD achieves its effects by entering into special "receptor sites" in certain brain cells. These cells normally take in serotonin through these receptor sites, but when these sites are blocked by LSD, the brain is effectively deprived of its serotonin. So far, therefore, all that we really know is that minute quantities of serotonin affect mental states, alter perceptions, and that new dimensions of conventional reality accompany changes in the level of serotonin in the brain.

We know that serotonin is produced in the brain by the pineal gland. We know that when its level in the brain cells is altered by the administration of minute quantities of drugs such as LSD, the perception of what we normals call reality is drastically altered.

To recapitulate briefly this patchwork account of various disparate facts and demonstrations, it is obvious that the pineal organ is a vestigial eye. What functional use any creature—fish, reptile, or mammal—would make of a third eye is still unclear; but it would seem that in those animals where the organ operates clearly as a light-sensor, it seems to provide some kind of metabolic regulatory function.

In mammals the pineal's most important role in the economy of the body is that of an endocrine gland. So far two hormones secreted by this gland have been identified. One of them, melatonin, appears to be an important regulator of sexual cycles in lower animals and a regulator of sexual maturity in humans.

The other hormone, from which melatonin is derived in the pineal through the action of an enzyme upon its molecular structure, is serotonin. Elsewhere in the body serotonin appears to promote blood clotting and muscle contraction, but in the brain it appears crucial to what is conventionally considered "rational" thought. If brain cells are deprived of serotonin, there results a disruption of rational thought.

In plants and in lower animals environmental light is a crucial factor in the regulation of sexuality, both physiological and behavioral. Virginia Fiske and Richard Wurtman had demonstrated a connection between ovary size in female rats and pineal gland weight. They were able to alter the size of both organs by altering environmental light.

It was this problem, the role of environmental light acting upon a vestigial eye, that began to intrigue Richard Wurtman. Working now with Julius Axelrod, the biochemist who had discovered how the pineal synthesized melatonin, Wurtman attacked the problem of light and the third eye.

It was possible that light entered directly into the skull. A California physiologist, William F. Ganong, had implanted photoelectric cells deep inside the skulls of various mammals and demonstrated that sunlight penetrated through skin and bone and brain in measurable and therefore significant amounts. Perhaps the eyelike structure of the pineal organ in reptiles still played an eyelike part in mammalian physiology as well, and served as a direct light sensor.[1] Or perhaps the central nervous system became aware of external light conditions via the conventional eyes, and it was this awareness which affected pineal functioning.

The pineal of the rat is very small—about one milligram in weight—and measuring subtle changes of enzyme levels within this minute organ was a difficult procedure. Wurtman and Axelrod were forced to devise techniques capable of great accuracy in order to measure these changes. But measure them they did. They wrote recently: "When rats were subjected to constant light for as short a period as a day or two, the rate of melatonin synthesis in their pineals fell to as little as a fifth of that in animals kept in continuous darkness."

They began exploring the role of light in this synthesizing activity by investigating the second hypothesis—that the conventional eyes served as light sensors, rather than the pineal gland itself. To this end they blinded rats and discovered that they "completely lost the capacity to respond to light with changed [synthesizing] activity; hence light had to be perceived first by the retina and was not acting directly on the pineal." The informational pathway, however, was via the central nervous system—the same feedback network that contracts or expands the pupils of the eye as light intensity increases or decreases.

Later experiments by Wurtman and Axelrod, far more complicated and extensive, seem almost anticlimactic. As their

[1] By means of a beautifully designed experiment, two German physiologists, Eberhard Dodt and Ewald Heerd demonstrated that in frogs, at least, the pineal eye still serves as a wavelength discriminator—it reacts differently to different wavelengths (or colors) of light.

earlier work had suggested, melatonin production in rats is closely tied to environmental light. Melatonin is produced in darkness, and production ceases during light. But in addition to enzyme activity the weight of the gland itself also changed, alternating in weight as did the conditions of darkness and light. Wurtman and Axelrod write that this suggested to them that "light was affecting many more compounds in the pineal than the single enzyme we were measuring." They believe (but are cautious about saying so) that there is an excellent likelihood of additional, as yet undiscovered, compounds being manufactured in the pineal gland.

The English word *ecstasy* derives from the Greek and means literally to stand outside; it also means to derange, to put out of place. In modern usage the word is commonly employed to describe a rapturous state of being in connection with either sex or religion. The emotional connection between the raptures of sex and religion is as old as mankind. It was only recently that Freud made the connection causal, and some of his more literal-minded followers determined that religion constituted a "sublimation" of sex, which implied that of the two, only sex was "real" and religion was an intellectual perversion of a basic primordial drive. Yet as discoveries in connection with the pineal gland proceed, the Freudian version of "reality" begins to differ from that reality perceived by the biochemists. Unlike Freud's, the reality of the biochemists is totally objective. It is becoming more and more difficult to avoid concluding that if ecstasy has any material biochemical basis in being, the biochemical substances controlling both its sexual and transcendental manifestations are probably manufactured in the pineal gland.

Strangely enough this possibility had struck Hindu philosophers certainly as long ago as the beginning of the Christian era. According to the Hindu system of Kundilini Yoga the site of the sixth or highest Chakra is located *in the physical body of man* in the "space between the eyebrows." As I understand it, a Chakra is a combination state of body-and-mind; neither can be separated from the other completely. The following verses of an old (A.D. 1526) Sanskrit manuscript attempt to describe this state of being,

"Here is the excellent (supreme) sixteenth *Kala* of the Moon. She is pure and resembles in color, the young sun. . . . From her whose source is the *Brahman*, flows copiously the continuous stream of nectar.

"Inside it [the pineal?] is *Nirvana-Kala* more excellent than the excellent.

"She is the ever existent *Bhagavati* who is the *Devata* who pervades all beings. She grants divine knowledge and is as lustrous as the light of all suns shining at one and the same time.

"Within its [the pineal's?] middle space shines the Supreme and Primordial *Nirvana-Sakti*; She is lustrous like ten million suns, and is the Mother of the three worlds. . . . She contains within her the constantly flowing stream of gladness and is the life of all beings. She graciously carries the knowledge of the Truth to the mind of the sages."

The verses quoted above from the *Sat-Cakra-Nirupana* of Purnananda, a sixteenth-century Yogi monk living in Assam, stress the brilliance of the unearthly radiance which seems, from the Western mystical literature as well, to be an essential part of the religious ecstatic state. The light is described above as being lustrous as ten million suns, but in another part of the poem he writes that it is also "soft like ten million lightning flashes."

Light—the lovely and mysterious effects and sensations of light pervade both the religious mystical literature and the scientific. Light and the Third Eye, the puzzling connection persists.

The great holy days of all religions tend to cluster around the dates of the solstices and equinoxes, those times of the year when natural environmental light alters its intensity and duration. Is it possible that certain susceptible individuals are more likely at those light-changing times, through altered pineal gland secretions, to experience transcendental, or mystical states? Do these holy days signify a cultural-historical recognition of this increased incidence?

Is the pineal gland in part responsible for "spring fever" and the springtime surge of human sexuality?

These recent discoveries—connecting the biochemical responses of the body to states of mind and to such subtle factors in the environment as changing light conditions—suggest that we may discover still more, as Bünning termed them, "cosmic happenings"—in addition to the rotation of the earth around the sun—which affect all living organisms, humans included.

There may be forces (and the peculiar and controversial ESP data would support such suspicions) acting upon the psyche of the individual and the collective psyche of mankind of which we have as yet only the vaguest glimmerings. That such indirect forces as the existence or nonexistence of environmental light, and the subtle passage of time itself exert tremendous pressures on the organism, we know without any question. We know that the experiences of subjects who have been dosed with infinitesimal quantities of LSD have shaken them to the very core of their being. We know that LSD operates via the natural chemicals of the body, and that the producer of these tremendously powerful chemical agents is the pineal gland. We know now without any doubt that the pineal gland is delicately responsive to those cosmic energy radiations we call visible light. Are there other emanations from the cosmos to which the pineal glands of certain exceptional individuals are also responsive?

And finally, do these forces have an effect via these exceptional individuals on that continuum of human transformations which we call the historical process?

Those ancient meditators on the human condition, who anticipated much of this modern scientific work on the pineal gland, have answered unhesitatingly "yes" to these questions. At this time the insights of the ancients and primitives as expressed in their writings and rituals lack credibility for the modern imagination. For us credibility is obtained through the exercise of the scientific method. But as these investigations proceed, it may yet come to pass that we shall find a new vocabulary through which we can structure our political and social ideologies so as to restate for our time that ancient consensus that the environment surrounding mankind is infinitely larger and more complex than current sociological and psychological dogmas hold; that in some nearly hidden corner of our inner selves we are aware of happenings elsewhere in the solar system and perhaps in the universe at large. We must find a way, as ancients did, and as primitives still do, to make this experience credible, and an intimate, meaningful part of the acts of our daily lives.

[3]

The Molecular Memory

We are transformed by time through living within it. We undergo tremendous alterations of form within our mothers' wombs, becoming metamorphosed from the zygote, a pair of fused single cells, to the billion-celled baby that emerges nine months later. And during the first trimester of life we are transformed from childhood into adulthood. In the final trimester of life we are again transformed until finally senescence and death overtake us. All these transformations exist within personally experienced time—are triggered by its passage.

We have also been transformed by another dimension of time —a far larger span of it—as has every other living organism abroad on the earth today. Within the hollow vastness of this almost unimaginable reach of time, the moment consists of the generation. This is the smallest cycle of this series of transformations, the alternation of life and death, and the metamorphoses that occur do so as parents beget offspring, whether the parent be a plant, a single-celled paramecium, or multicelled human being.

Each generation has incorporated within it a minute element of transformation from its parent. There are frightful gaps in our understanding of the historical sequence of these transformations. The bony, fossil record is incomplete, and even over those fragmentary evidential scraps that we possess, there rage continuously the most fervent academic controversies. Though they may be conducted by academics, and technically beyond the reach of a layman's competence for judgment, they are far from unimportant. They involve us all in questions of racial lineage. Hopefully the problem will soon become academic in the other sense as well, and become socially unimportant as mankind gradually becomes one huge interbreeding population. Now man has reached that state which in his vanity he believes to be sapient. As a species he must recognize that he is bound together far more by his separation from other forms of terrestrial life, than he is alienated by any taxonomic differences between the various members of his own extended species.

These odd and inexplicable discontinuities in the fossil record present us with mysteries concerning the pace of change, and even more importantly, the purpose of change. Has evolution a purpose? Science considers itself unfit to respond to this question, but there is one teleological fact which intrudes itself so insistently that even science has been happy to acknowledge it. Evolution has, if not purpose, at least direction. It proceeds from states of structural and behavioral simplicity to states of structural and behavioral complexity. Obviously simple as this statement seems offhand to be, it nevertheless presents a terrible paradox to science. As a process evolution defies (or contravenes) the second law of thermodynamics, which proposes that all known systems within the universe proceed in the opposite direction, toward ever greater randomness, toward chaos, not order. Of course, as a system evolution cannot be considered closed; the very vastness of its scope opens it up to all manner of energy influences. Yet traces of the process as a whole exist within each of us as individual closed systems. As the individual matures from embryo to adult, structural traces of previous states of evolutionary being may be discerned. Memory of these former states is exhibited in structure. For each of us then, as we pass from nonbeing into being, there moves with us like the tail of a comet, as we enter into our perihelion of the present, remembrances of infinitudes of time-past. This is memory of a strange sort—memory manifested in biochemical energy.

This strange memory is manifested in each of us, and in every other organism abroad upon this planet; it is also manifested by each and every one of the billions of cells which compose that community of being we call human.   Energy is absorbed by each of these cells and transformed into behavior.   In most of our body cells this energy is transformed into the behavior of biochemical and electrical activity.   In the male there exists a community of free living cells, the spermatazoa, which after ejaculation, live active behavioral lives quite independent of the parent organism.

All of this behavior—whether it be covert and therefore difficult to observe and analyze, as when it consists of energy transformations by a cell, located say, in the liver; or whether it be overt as it is when we watch sperm whipping their long tails, swimming vigorously under the microscope—all this behavior depends on something that, for want of a better name, might simply be called memory.   Somehow the cell is rendered capable of re-enacting a prescribed behavioral sequence.   This memory is of an entirely different order from that which we commonly associate with the recollection of a fixed experience in the individual mind.   This chemical, or rather molecular, memory does not result from the experience of the individual.   It pre-exists in the organism.   It is innate.   It derives from other memories— from the memories of the dead, from memories that run on defying all conventional views of time.   This memory is part of the ever-moving present.

One may, in fact, quite accurately define life itself in terms of this memory.   Living systems are complex chemical arrangements possessed by memory.   Nonliving systems do not appear to have it.   At least not to the same degree.

For example, some of the viruses, as we understand them, occupy that borderline between living systems and inert matter, appearing to be little more than a crystallized protein of enormous molecular weight.   Like ordinary crystals they can multiply themselves—as ice, for example, may form across the surface of a lake—but unlike ice crystals, viruses are capable of change, of mutation.   They are alterable, by experience, in form and

function.   And this alteration is permanent, for as they reproduce
themselves, the change is transmitted to the newly formed
generation.   The carrier of this change would appear at this time
to be one of the nucleic acids, merely a mass of molecules principally composed of nitrogen and phosphorus—inert things.
Nitrogen is most frequently seen in gaseous form, and phosphorus
as a mineral.

As with the pineal gland, the function of which was intuited
by preliterate vedic peoples, so intuitions (expressed in ritual)
concerning the molecular memory have existed in humans for a
long time.   This intuition is far more widespread than the
particular one concerning the pineal.   In fact it seems to be
universal, and dates back into the Pleistocene era.   It shows itself
in the idea of the Resurrection.   Most of what little we know of
late Pleistocene man, we know because he buried his dead.   And
moreover, he often buried his dead with the furnishings of life,
as though he anticipated a resurrection of the corpse which
would then have need of these accouterments.

The general notion of the Resurrection, of life-after-death,
involves a constellation of separate ideas.   Among them is the
primordial fear of death and consequent extinction.   It *appears*
superficially as though death involves a total extinction.   But this
extinction involves with certainty only the person.   Curiously
enough, the word *person* derives from the Latin *persona*—
literally, through sound—and it is through sound,[1] through
language (including the analogous language of behavior), that
the personality expresses its uniqueness, becoming recognizably
unique to others.   All the accumulated experience of the living
with the dead tends to support the idea that the person is
extinguished in death.   But there is no hard evidence to contra-
vene the supposition that life-as-memory may yet persist in the
chemical molecules of the cells.   If so, it is quite possible that
experience does persist after the one who lived through the
experience is dead or extinct.

[1] As sound issues through a mask, for the Latin word *persona* means mask.

Though Darwin's later ideas on germ-cell memory have been discredited for the last hundred years, several distinguished biologists now consider seriously the possibility that the molecular memory of an experience may conceivably be transferred from brain cells to reproductive cells and thus to another generation

Actually, by substituting the notion of a persisting molecular memory located in the individual cell, for the notion of a resurrection of the entire corpus of a man, one is merely making the scale of the occurrence—the size of the organism involved—the determinant of credibility. On the cellular level we do not consider the miracle to be at all miraculous, in the sense that we consider miracles to be highly improbable events. This miracle occurs continuously in all of us, as inert materials gathered from the environment are transformed into compounds which become living matter possessed of memory. Thus is the mystery central-ized into its truly essential aspect: the metamorphosis of matter into energy and vice versa. As in the gross and macroscopic event of the ascent of man, science deems itself unfit to theorize about the teleological aspects of the microscopic event.

Yet it was a perfectly respectable scientist, engaged in perfectly respectable scientific procedures, who first discovered the event for our time in the name of science. His name is Fred Griffith and he was working in His Majesty's Ministry of Health in London during the 1920's. In the course of his work as a bacteri-ologist he had injected some mice with a nonvirulent form of *diploccocus pneumonia*. Pneumonia bacteria represent a large class of amoeba. There are nearly one hundred separate species so far identified. Some of them are virulent and some of them are not; this is an easy identification to make. The virulent bacteria appear under the microscope as small flattened pearls; they are encased within a smooth capsule of carbohydrate. The non-virulent types look more like what a layman would expect an amoeba to look like—an amorphous blob, a Joan Miró shape They lack the capsule.

Occasionally from within a population of smooth, virulent

bacteria a bloblike amoeba will appear.   The appearance of this individual is occasioned by mutation, in much the same way as an albino individual will appear occasionally within a population of normally pigmented persons.   The mutant, nonvirulent pneumonia bacterium can be isolated and induced to reproduce true to form.   In other words, virulence or nonvirulence is controlled by one or more of the bacterial genes.

Griffith first injected his mice with nonvirulent amoeba from the Type II strain of pnuemonia.   He then injected the mice with the dead bodies of virulent Type III bacteria.   He had killed these creatures with heat, putting a culture dish of the tiny, smooth, pearl-like animals in a sterilizing machine.   He prepared a solution containing these corpses and injected them into his mice. In due course, much to his astonishment, his mice became ill with pneumonia, and upon examining them, Griffith discovered that virulent Type III bacteria abounded in their systems.   His first thought was that these creatures had been resurrected—perhaps they had not been properly killed, but placed in a state resembling death, and they had "revived" in the hospitable environment of the bloodstream of the mice.   Being a careful and skeptical worker, he immediately excluded this possibility.   He performed the experiment again, making sure that the virulent bacteria were dead.   Again they appeared alive in his mice.   The dead virulent bacteria had seemingly been restored to life.   Or there was another possibility: the living nonvirulent bacteria had been transformed into the likeness of the dead merely by proximity. Either way, it was an unheard-of occurrence.

Griffith published a report of his experiments in the British *Journal of Hygiene*, in 1928.   Unfortunately, it received little attention outside of the biological community.   Before the year was out, and before the experiment could receive widespread acceptance—before, that is, it could be repeated in other laboratories by other workers—the Great Stock Market Crash had occurred in the United States; and all over Europe erupting social and political turmoil usurped the attention of those laymen and humanists who might otherwise have reacted to this occurrence as being possibly as far reaching in its philosophical implications as was Darwin's hypothesis. The medical biologists, however, seized upon the possible practical applications of Griffith's discovery and many modern immunization procedures derive from a technical understanding of this principle.

Another reason why this event went relatively unnoticed in the wider world is that laymen generally see no necessary basis for comparison between the genetic systems of bacteria and the genetic systems of human beings. Yet as systems they are closely analogous, and since Griffith's time some of the most fascinating work done in genetics has been done with microorganisms. Geneticists have always had the problem of being required to rely on occasional mutants, which may seldom arise from amidst large populations of animals. In order to obtain any kind of statistical accuracy, very large numbers of animals must be involved in the experiments. Also, since the lifespan of the observer is limited, these animals should optimally enact their cycle of death and rebirth within a comparatively short span of time, so that a significant series of generations may be observed. When microorganisms are used, populations numbering millions of animals can be maintained in very small quarters, and their life cycle completed in sometimes a matter of hours. Before procedures became sufficiently sophisticated so that microorganisms could be widely used (since the turn of the century), fruit flies were the favored creatures of geneticists, and genetic findings applicable to these creatures were directly translated as being pertinent to the human condition.

Within the next sixteen years from 1928 to 1944 innumerable investigators reproduced Griffith's work. During this time, it became obvious that what was happening here was not a resurrection but a transformation. The two ideas are very closely associated. Analytical psychology sees the resurrection as a symbolic death and rebirth occurring within the individual psyche. According to this view they are not literal occurrences.

Neither the individual or his psyche actually dies. The psyche becomes transformed, metamorphosed into something other and different than what it was. This sophisticated recognition of the symbolical nature of resurrection, which doubtless underlies the ceremony in whatever cultures it appears, need not necessarily be held by the mass of communicants. They probably always tend to view the resurrection as being a far more literal affair, but also one involving a transformation—a transport of the person into another realm such as "heaven" which would necessarily involve a total reorganization of the individual. It is highly unlikely that ancient man literally believed, any more than we do, that the dead are likely to be resurrected in *this* life.

But Griffith's experiment proved that the phenomenon *does* occur in actual fact, that by their presence among the living, the dead had altered the living, created among them a metamorphosis. Were the myth of resurrection merely a symbolic representation of change, it would hardly need to be so elaborate and unbelievable. It is more than merely "change," it is testimony to the fact that experience does not vanish as the incident that produces the experience terminates. The incident persists in memory. But even more importantly, the memory is projected. The capacity for speech, for example, is now innate in humans. The memories of well over a hundred thousand years of speech have been coded into the molecules which are inherent in all the cells of the body. This is the resurrection: the fact that the dead move among us the living, with their accomplishments alive in our flesh.

We are accustomed to viewing this effect of the dead upon the living as being metaphysical, a cultural influence. It is customary to credit the transformation of the living by the dead as accomplished *solely* through tradition and the artifacts of culture. But can one credit the persistence of archetypal memories as expressed in the relatedness of myths and folktales or even in the raw fact of speech itself, solely to cultural effect? One can expose an intelligent chimpanzee to all the culture and tradition imaginable, but will he speak?   Or even if he does, will his offspring begin to speak spontaneously during their later infancy?   Not very likely —for the dead must be resurrected again and again within and among the living for such transformations to build the capacity for speech.   Whole neutral networks capable of creating abstract symbols do not appear within a single generation.   They must be evolved. And some glimmerings of insight into the mechanics of how such an evolution is possible may be gained from an examination of this and other related experiments.

In Griffith's experiment the effect of the dead upon the living was direct and immediate.   It was also unmistakably structural. The dead had altered the living into a likeness of their dead selves.   Sixteen years had to pass from the date of Griffith's publication of his work, and millions and millions of bacteria had to be transformed from one physical entity into another, before the next step could be taken.   During all this time it had never been possible to remove the transformation from within the blood and tissue of host animals, mice, rats, or guinea pigs. Before the next step could be taken, and the whole phenomenon analyzed objectively, transformation would have to be made to occur *in vitro*, in glass, outside the body where it could be observed as an ongoing process.   It was only in 1944 that this was finally accomplished by three physicians, Oswald T. Avery, Colin McLeod, and Maclyn McCarty, at the Rockefeller Institute. They managed to create for the occurrence a hospitable artificial medium.

Originally Avery and his associates suspected that the carbo-hydrate capsule surrounding the virulent bacteria—the armored sheath that gave the creature its distinct appearance—was involved in the transformation.   Once they were able to reproduce this transformation occurrence in glass, they began preparing extracts from the corpses of the dead and introducing them into the company of the living to observe the results.   In this way they were able to eliminate factor after factor, finally arriving at that substance which caused the transformation.   In the end it turned out to have nothing whatever to do with carbohydrate. It was a far more complex molecule, and almost immediately it proved to be an infinitely more universal and fundamental material than the carbohydrate shell would have been. The material was a complex chemical structure with the forbidding name of deoxyribonucleic acid, later abbreviated to DNA.

Immediately following on Avery and his associates' success in grappling with the molecule concerned with this transformation, a host of investigators working within a dozen different disciplines attacked the acid, hoping to find within it the clue to inherited memory. In 1953 James Watson, now of Harvard, and Francis Crick of Cambridge, England, contrived a "theoretical model" of the molecule of this acid, for which they belatedly received the Nobel Prize in 1962. It took nine years for the world-at-large to realize the tremendous importance of these obscure happenings in obscure creatures. Today the whole event has been absorbed into the imagination of those who must work with it daily; it has become a commonplace. For example, Robert P. Levine, the distinguished Harvard geneticist, can now write flatly in an introductory genetics text: ". . . transformation entails the exposure of bacteria of one genotype to a solution of highly purified DNA derived from the bacteria of another genotype. The recipient bacteria take up this DNA, it becomes integrated into their genetic material, and ultimately some of them express the genetic characteristics of the donor bacteria."

DNA is a neucleic acid, and nucleic acids are chemical chains made up of links called nucleotides. Each nucleotide link is itself a complex molecule consisting of three basic components: a nitrogen base, a sugar, and a phosphoric acid radical. So far as we know, there are two main classes of nucleic acids; the deoxyribonucleic acids (the DNA's) and the ribonucleic acids (the RNA's). Chemically they differ from one another in their sugar content and in their nitrogen bases.

Their physical properties are quite different.

In the living cell, at least of higher organisms, almost all the

DNA is located in the nucleus, the central enclosure within the cell that contains the chromosomes. Some RNA is also found here within the nucleus, along with traces of protein, but most of the RNA found in the cell is found in its cytoplasm or flesh. DNA responds to the Feulgen stain by which the chromosomes are known—their very name means, literally, the *colored bodies*— and it is the DNA which gives them their color under certain conditions. The primary messenger, then, of genetic information is the DNA.

Beginning with Gregor Mendel's work, through the investigations around the turn of the century by Thomas Hunt Morgan and others, the chromosomal theory of inheritance gradually became established. This hypothesis claimed that the *total* heredity of the cell, the sum of its molecular memory, could be accounted for by elements that were called genes. These were supposed to be strung along the length of the threadlike chromosomes in the same way as pearls in a necklace.

The genetic information contained in the DNA, which comprises the chromosomes, seems to be relatively fixed and unalterable; only certain mutagenic agents (X-rays for example) can penetrate the nucleus and alter the chromosomes. Once a hereditary character becomes embedded on the chromosome it is not easily susceptible to change. And when they do occur, such changes are more likely than not to be deleterious or even lethal. According to the chromosomal theory of inheritance, which became incorporated into the theory of evolution, all the hereditary factors were located on the chromosomes, and these were only alterable by totally random chance occurrences, such as the chromosomes being struck by a cosmic ray, and the remote possibility that such a random change would be benign and give the creature thus changed an adaptive advantage.

Griffith's work disclosed the possibility that alterations in genotype could come about by other than purely random happenings. At the time he was doing his work, the evidence was overwhelming in support of the chromosome theorists' contention that all the significant *structural* characteristics of the cell—or the organism resulting from the community of cells—was controlled by elements within the nucleus. But there are other characteristics aside from the structural; the migration route of the lesser whitethroat for example, in all the magnificent detail of its direction-indicating guidance system, is also an inheritable trait.

The term gene—which forms the root word for its successor concept, genotype—is, like its derivative concept, also metaphysical. When one transposes the notion of the memory of an exhaustive journey such as the migration of the lesser whitethroat into the concept of the gene, one becomes lost in an intellectual swamp. And the evidence that such inherited memories are carried on chromosomal genes is not at all clear-cut. On the contrary, there is considerable evidence that behavioral characters are not carried on chromosomal genes. And there is, as every horse player knows, no necessary correlation between a thoroughbred's physical configuration and those intangibles of "heart" which make him a winner at the track. Man has known for eons that character traits are inheritable; he has bred innumerable servitor animals for special purposes dependent upon the development of special character or behavioral traits. There are dogs bred for sheep and cattle herding, for retrieving, for tracking and trailing; there are pigeons bred for their "homing instinct," and bulls for their bravery in the *corrida*, and so on.

The fact that animals can be bred for them, indicates that there is a sizable genetic component in these behavioral traits. But it is still very difficult to locate precisely the site of those genes which control behavioral traits. Are they in the chromosomes? Possibly—especially in those instances where the trait is stable and associated with structural type. Chromosomal genes *are* stable. Cytoplasmic genes are not necessarily so—they may be stable enough in the lifespan of an individual, but they may also appear and disappear within a pedigree in unpredictable fashion.

Evidence that the chromosomes did not carry the total

hereditary memory of an organism had appeared even before the turn of the century. But the chromosome theory "worked," for most structural characters and proponents of the theory were able to kick evidence of another system under the rug, though in so doing they were often forced into courses of conduct more becoming to dogmatic ecclesiastics than to supposedly open-minded, objective scientists. Intradisciplinary geneticists' quarrels, even as they appear in the published literature, take on the tone of medieval scholastics arguing the number of angels who may dance on the head of a pin. And in fact, such disputations concerning the size, structure, and appearance of genes, and where they dance, may well be extensions of these medieval debates. When these scholastics spoke of angels, they may well have been speaking of what today are called genes. For example, Moses Maimonides the Jewish philosopher writing in the twelfth century, during the heyday of controversy concerning the structure and appearance of angels, grew bored with it all. " 'Angel,' " he wrote, in strikingly contemporary language, "means messenger, therefore every one who is entrusted with a certain mission is an angel. . . . Say to a person who is believed to belong to the wise men of Israel that the Almighty sends His angel to enter the womb of a woman and to form there a fetus, he will be satisfied with the account, he will believe it . . . although he believes that the angel consists of burning fire and is as big as a third part of the Universe, yet he considers it possible as a divine miracle. But tell him that God gave the seed a formative power which produces and shapes the limbs and that this power is called an 'angel' . . . and he will turn away."

This problem of what Maimonides called the "formative power" is as crucial today as it was then, and those corollary discussions of just what an angel or a gene is and where it may dance are equally crucial. They involve our understanding of the nature and permanence of experience. Do the experiential memories of many individuals accumulate and become part of the heritage of the species, or do they decay as the flesh of the individuals who experienced them decay? Do these experiences participate in giving direction and possibly purpose to evolution, or is mutation a totally random process? It is possible that answers to some of these basic questions of existence may be found in the study of cytoplasmic genes.

In their book *Cell Heredity* Ruth Sager and Francis Ryan offer a definition of a gene which rivals Maimonides' in its humility, in its sense of awe and wonderment. They describe the gene as "a hereditary determinant which, in its alternative forms is responsible for differences in a particular trait. We do not specify either its location in the cell (chromosomal, non-chromosomal) or its constitution (DNA, RNA, protein), because flexibility and generality are necessary in the face of rapid advances being made in this field. . . ."

This modest and open-ended definition was offered to the profession in 1961. Earlier than that, Sager and Ryan would have had a great deal of difficulty getting such heretical vagueness to be generally accepted. To remain a respected member of the profession, one did not shrug off the chromosomal catechism so easily, in print. Until very recently the profession of geneticist was truly a discipline in every sense of the word.

As much as any one man it was a stubborn, painstaking geneticist named Tracy Sonneborn, working at the University of Indiana, who by the cumulative efforts of his life's work literally forced open the closed minds of his professional peers. His demonstrations were so elegantly conclusive and so easily reproduced by those who wished to see for themselves, that his discipline has not been quite the same since he made his appearance within it.

One day in the early 1940's Sonneborn went fishing for paramecia in an Indiana pond, with his net and Mason jar. He wanted some wild animals taken from nature to refresh his laboratory colony, which had been interbreeding for hundreds of generations. He found some wild animals—they were so wild that soon after they were placed in a container holding lab animals, the latter died. This was curious.

Paramecia are small, single-celled animals shaped roughly like a comma, and about half the size of a comma in five point type, the size of type usually found in unabridged dictionaries. They are classed in the order of protozoa, a coinage from the Greek meaning, literally, first living beings. Some paramecia species ingest their food by surrounding their food and absorbing it in the manner of amoeba; some possess a primitive type of mouth or gullet. The outer surface of the body is protected by a skinlike membrane known as the pellicle. Within this pellicle is the cytoplasm or flesh of the animal, which has roughly the consistency of cool molasses. Being a single-celled creature, it does not possess internal organs as do multi-celled creatures. Instead there are, embedded in the cytoplasm, small pulsing sacs called vacuoles, which perform the various necessary tasks in connection with ingestion of food, respiration—the regulation of body chemistry. Growing out from the pellicle in most species are numerous hairlike cilia. By moving these in rhythmic fashion, like the banks of oars in a trireme, the animal can swim forward or backward in its liquid medium.

The sense receptor system is imperfectly understood, but the animals respond behaviorally to light, temperature, and chemical changes. They behave purposefully, being attracted to certain stimuli and being repelled by others. Some experimenters have claimed these animals are "trainable" by simple reward-punishment techniques, but this work is still controversial. In essence then, though it has the simple structure of a single cell, the paramecium is nonetheless a complicated animal. They have a primitive courtship ritual—that is, a behavioral sequence which invariably precedes mating—and one of their closely related species, the *Hypotrichida*, have evolved this courtship ritual to such a point that it is called, even in the genetics literature, a mating dance.

At this low level of structural organization, genetic exchanges between partners is not called mating, but conjugation. It is quite other than the sexual matings of multi-celled organisms, in that no zygote is formed, the zygote being the fertilized egg which produces a community of cells out of which a larva or fetus will form. Instead, paramecia reproduce by simple fission. What happens is this: The DNA, which comprises the basic stuff of which the chromosomes are made, is located in one or more sacs called nuclei. The DNA duplicates itself. This is an inevitable precursor to all cellular duplication—whether the cell involved be a single cell duplicating itself as an individual entity, as in the case of the paramecium; or whether it be a single cell as part of a community of cells, as in the case of human growth or the healing of injuries, etc.

The mechanics underlying this process of DNA reduplication are being examined today in every nation with the techology to do so. But it still remains, as of this writing, a mystery. Whatever the mechanics may be, the process is accomplished, the DNA reduplicates itself, and the sacs containing the DNA migrate to opposite ends of the animal. Then the animal itself forms into a new shape, a kind of dumbbell or hourglass shape, forming a small waist, which narrows until finally it parts and there are two organisms where formerly there had been but one. This reproduction by simple fission is known, sensibly enough, as autogamy, literally marriage to oneself. It was formerly believed that the protozoa genotype of the individual was "immortal" in that the individual animal could perform autogamy an infinite number of times. Now we know this not to be the case. There are only a certain number of autogamous reproductions possible for any species of paramecia. After several hundred autogamous reproductions there will be ever-increasing time lapses between reproductions until at last the creature will perish without an heir. Toward the end of this chain of generations the animal seems impelled toward conjugation. Paramecia can also be induced to conjugate if an ample food supply for the community is suddenly depleted, or in some cases by chemical or temperature alterations of the medium.

What happens in conjugation is this: In each prospective mate, prior to the actual act of conjugation, the DNA reduplicates itself within the nucleic sacs. However, unlike the similar happening during autogamy, only half the amount of DNA appears in each sac; it is, in the language of genetics, a haploid nucleus, containing only half the required number of chromosomes. The human egg and sperm also contain half the amount of DNA that any other body cell does. To fulfill their predetermined behavioral cycle, each cell must bring to the other the missing half of the DNA requirements. The two paramecia, now each containing haploid nuclei, swim close together touching body surfaces; the nuclei have migrated close to the pellicle, and then through their various membranes as if through a sieve, each animal loses one of its pair of haploid nuclei to the other. It now possesses the full complement of DNA material, half of it a reduplication of its own body-materials, and half of it derived from its mate. At this point both animals undergo a regular fission process, forming the hourglass shape and then separating.

Occasionally paramecia will exchange more of one another than merely nucleic materials. Instead of merely touching pellicles at the bulge end of their comma bodies, they will also come together at their forward end as well. In this way elements of cytoplasm or flesh are also exchanged—they stream from one animal to the other. Under normal circumstances, if paramecia are left alone to accomplish this spontaneously, this cytoplasmic conjugation is a comparatively rare occurrence. But the probability of cytoplasmic conjugation can be greatly increased by special procedures altering light, temperature, chemistry of the medium, etc.

Sonneborn was fascinated with his new lethal paramecia; what was it about them which caused his older, laboratory animals to perish within a few hours, when they were both placed in a common medium. After some study he discovered that his new, as he called them, "killer strain" secreted a substance into the medium which he named paramecin; this paramecin was lethal to his laboratory animals. Obviously animals possessed of this poisonous armament would have a survival advantage over other animals. Sonneborn was curious how the trait was transmitted. Getting sensitive animals to conjugate with killers and still survive presented certain problems. Sonneborn watched his communities carefully, waiting till two animals were preparing to conjugate with partners from their own population. Then, just at that crucial point when they were about to swim together, he would transfer them by pipette, each adjacent to another, and let them perform their conjugal act, separating them afterwards before the "sensitives" could succumb to the paramecin.

He noticed immediately that the killer trait was never transmitted during conventional conjugation when only nucleic materials were exchanged. There had to be cytoplasmic conjugation before the killer trait would be likely to appear in subsequent generations.

After more painstaking work Sonneborn discovered that the paramecin was secreted by small elements, in the cytoplasm of the killer animals, which he called Kappa particles. It was these particles that had to be transmitted during cytoplasmic exchange; but once transmitted, they became part of the genotype of the animals receiving them. Their offspring would then be possessed of the trait and could, moreover, pass it on by conjugation, in their turn, to still other animals. Here at last was an angel that could be seen! The Kappa particles were comparatively large objects, larger than most viruses but smaller than most bacteria. The particle was composed of DNA and protein. Sonneborn published his findings in 1943, and naturally chromosome theorists postulated that the Kappa particles were an independent organism which had evolved a kind of parasitic relationship with its host, the paramecium. They were, it was claimed, a kind of infectious agent, and the transmission of this agent, while seemingly hereditary, was really nothing more than the transmission of an infection. Their contention was strengthened by the discovery that Kappa particles multiplied at differing rates from their hosts. The particles could themselves be isolated from killer animals and introduced into sensitives, converting the latter into killers.

But in the years following, subsequent research showed that there was an interaction between the Kappa particles and at least one chromosomal gene; animals possessed of the particles, but not possessed of the gene, could not maintain the particles—if introduced, they would, over a period of time, disappear from the animal's cytoplasm, and it would revert to being a sensitive. It seems now as if some element from the Kappa particles enters into the nucleus and attaches itself to the chromosome like a passenger boarding a train. The important discovery that Sonneborn made was that, even though some kind of alteration was made to a chromosome, this alteration was particular and distinct and predictable. And most importantly, no transfer of nuclear DNA was involved in acquisition of this alteration.

It depended, however, on still further work that had to be done before the doctrine of cytoplasmic genes could become widely accepted. Many experiments were then performed with disease elements that attack bacteria. These are viruses known as bacteriophages. Almost invariably, the effects of their attack are mutagenic, roughly analogous to the effects of German measles on human mothers. The existence and importance of nonchromosomal (or cytoplasmic) genes is therefore no longer seriously questioned, at least as regards its importance to disease.

But more recently Ruth Sager and her students at Columbia University have discovered another concept-shattering event. Sager's experiment discloses that the molecular memory is not merely a passive storage unit of past experience. The memory molecule itself *responds actively to new experience*! And this new response is then coded into the molecule for the information and preservation of future generations. Sager chose for her work a bacteria known as *Chlamydomonas*, belonging to that extended family of halfway creatures (neither wholly animal, nor wholly vegetable) the green algae, which normally inhabit stagnant ponds in summer.

She placed her bacteria in a lethal environment—a medium containing streptomycin.   When many of these so-called wonder drugs were first employed, they were 100 per cent effective, but the evolutionary process gradually responded to this new environmental threat and drug resistant bacterial strains began to appear. Sager was interested in the genetic mechanism underlying this response.   Traditional chromosome theory held that the surviving organisms possessed a pre-existing trait giving them immunity, and that this trait had been conferred via a random mutation to a chromosomal gene—the result perhaps of a cosmic particle originating in the sun, which shot through space, through the earth's atmosphere, finally striking the right spot on the microscopic chromosome of a bacteria, deforming the chromosome, and thus producing the trait of resistance.   Sager was interested in finding out if this somewhat simplistic explanation adequately described the whole process.

After leaving her bacteria in the streptomycin for a reasonable period, Sager inspected her cultures and found that almost all of her bacteria had died.   Only one bacteria out of each million survived exposure to the drug.   "Traditional genetic procedures," Sager writes, "disclosed that most of these survivors were immune as a result of a mutation to a chromosomal gene. About 10 per cent of these survivors, however, had an immunity trait caused by a nonchromosomal gene.

"We found," she continues, "that the chromosomal mutations were spontaneous, arising at random before the cells were placed in a streptomycin-containing medium. . . .   So far this conformed to theory.   The nonchromosomal mutations, however, arose only after sensitive cells had been grown in the presence of toxic but sublethal concentrations of streptomycin. They were induced mutations."

That last simple four-word sentence may, if her findings are further confirmed and the mechanism better understood, require an entire revision of evolutionary theory.

Sager continues, stating that she and her co-workers found that "the streptomycin was acting nonspecifically, and that it could induce mutations in many nonchromosomal genes. . . . This remarkably high level of mutagenic efficiency is quite different from that of traditional chromosomal gene mutations, which affect only a small fraction of the population."

Her words speak for themselves.   In order to produce genetic alterations of an organism, she placed them in an environment which was dangerous but not lethal.   Given, thus, an opportunity to respond—but more importantly to survive and thus pass on the response—the molecular memory inherent in the flesh, the cytoplasm, responded.   It did not respond specifically, but this does not matter.   If we extrapolate the occurrence by using a human analogy, we too respond to threat variously and non-specifically.   Our cerebral intelligent response to a threatening movement in front of our faces is equally as various as the non-cerebral molecular response of the *Chlamydomonas* bacteria is to its threat.   We humans will probably do several different things: We will blink our eyes, pull our heads back, raise our hands protectively, and very likely give a warning cry of some sort.

Important as its theoretical implications are, Ruth Sager's work still involved microorganisms whose behavior patterns are so indistinct as to be far removed from anything that we humans can anthropomorphically construe as "learned behavior."

The central puzzle which has obsessed psychological theorists for the past several hundred years, and philosophers long before that, resolves itself in the word *instinct*.   What is an instinct and how is it acquired?   The foregoing reports of Sonneborn's and Sager's experiments are intended to propose that perhaps cyto-plasmic genes—being relatively unstable as compared to chromo-somal genes, and therefore responsive to events connected with the environment of the organism—may be involved in the evolution of what is loosely called instinct.   The English word itself derives from the past participle of the Latin word meaning "to incite." That is as good an explanation of what the mechanism very well may be, as any other. An experience must be sufficiently jarring to *incite* the molecular memory to respond.

In 1956 two graduate students in psychology in a University of Texas laboratory stumbled over another incredible occurrence. They were trying to prove that worms could be taught, or made to learn, and had chosen, not the garden variety angleworm, but one of the flatworms of the *Platyhelminthes* family. These represent one of the major phyla of the animal kingdom. Most of the members of it are parasitic: they include tapeworms, liver flukes, and so on. Many species have become extraordinarily specialized: one has been identified that seems able to exist happily only in the eye of a hippopotamus. They are a curious and primitive form of life.

Theories concerning their place in evolutionary history abound. They appear to be one of the first, if not the very first, of bilateral forms. Such creatures as jellyfish and sea anemonies are not bilateral. But beginning with starfish and ending with humans, most other creatures, including all the insects, are bilateral—one side of the body being a mirror image of the other. And very likely it all began with the *Platyhelminthes*. They do not possess a proper brain as we know it, but they do have a snarled complex of nerve endings pulled together at one end of their body, which end shows signs of being an evolutionary precursor of a head. In the Planaria worms which these two graduate students James V. McConnell and Robert Thompson used for study, this headlike end of the worm is flared like an arrow point and has two light-sensitive patches to either side, which look like the possible evolutionary precursor of an eye. They possess a synaptic nervous system like that of human beings, but they have one great facility we lack—they are capable of regeneration. Cut a planarian in half, the tail will grow a new head, and the head will grow a new tail. Sometimes a half-inch planarian can be cut carefully into as many as fifty separate pieces without destroying it, and each piece will regenerate itself as a complete worm.

McConnell and Thompson began their work by shining a bright light on the worm at the same time they gave it a mild electric shock. When it felt the shock, the worm convulsed, and McConnell and Thompson were curious to know if they could create in the brutish "mind" of this creature, an association between light and shock, so that when they shone the light alone, the worm would convulse as if "expecting" the shock to follow. If they succeeded in this, it would be the first time that true "learning" had been demonstrated in so lowly a creature. They did demonstrate that the worms could acquire a traditional Pavlovian "conditioned reflex." Not only did they teach this simple accomplishment, but in the years to follow, McConnell and his associates have "taught" worms such difficult exploits as keeping to the white squares in a checkerboard pattern for fear of shock punishment.[1]

McConnell himself describes what followed in their work: "It was while we were running the first experiment that Thompson and I wondered aloud, feeling rather foolish as we did so, what would happen if we conditioned a flatworm, then cut it in two and let both halves regenerate. Which half would retain the memory? As it happened, we never got around to performing that experiment at Texas, for Thompson received his doctorate soon after we finished our first study and went on to

[1] It is only fair to report here that controversy continues to surround the McConnell experiments. In the early part of 1965 perhaps their most distinguished critic, the 1961 Nobel laureate in biophysics Dr. Melvin Calvin of the University of California, took the field. Dr. Calvin attacked the McConnell experiments on the following basis: He claimed that for learning to be *true*, a substantial portion of any given population must be capable of acquiring it. As a criterion of learning he required an animal to respond correctly eighteen times out of twenty trials. But only nine of his fourteen trained worms reached this criterion. According to Calvin's definition of learning, therefore, 65 per cent of any given population is not a substantial number. This seems to me a capricious definition. It seems obvious that certain difficult skills, such as for example, playing Johann Sebastian Bach's *Goldberg Variations* on the piano, cannot be successfully taught to any substantial portion of the human population. But this does not mean that if a person acquires this skill he has not undergone a true learning experience.

Louisiana State University and bigger and better things—namely rats. When I went to the University of Michigan in 1956, however, I was faced with the difficult problem that in the academic world one must publish or perish. The only thing I knew much about was flatworms, so I talked two bright young students, Allan Jacobson and Daniel Kimble, into performing the obvious experiment on learning and regeneration. . . . In all honesty, I must admit that we did not obtain the results we had expected. We had assumed that the regenerated heads would show fairly complete retention of the response for, after all, the head section retained the primitive brain and 'everybody knows' that the brain is where memories are located. . . . We had also hoped in our heart of hearts, that perhaps the tails would show a slight but perhaps significant retention of some kind merely because we thought this would be an interesting finding. We were astounded, then, to discover that the tails not only showed as much retention as did the heads, but in many cases did much better than the heads, and showed absolutely no forgetting whatsoever. Obviously memory in the flatworm was being stored throughout the animal's body, and as additional proof of this we found that if we cut the worms into three or even more pieces, each section typically showed a clear-cut retention of the conditioned response. It was at this time that we first postulated our theory that conditioning caused some chemical change throughout the worm's body. . . ."

McConnell had published his data on learning in the regenerated halves in 1957, and two biochemists, Roy John and William Corning, working at the University of Rochester, became interested in the problem. "John reasoned," writes McConnell, "that learning in flatworms had to be mediated, in part, by some molecular change within the organism's cells. . . . John believed that RNA might be implicated in learning and retention in planarians. So he and Corning conditioned a number of flatworms, cut them in half and let them regenerate in a weak solution of ribonuclease [an enzyme] which breaks up RNA.

When they compared their experimental animals with a number of controls, they found evidence that the experimental heads were relatively unaffected by the ribonuclease, while the tails showed complete forgetting." McConnell interpreted this by imagining that memory is a twofold enterprise. It involves changes in the brain, *but also involves* "a change in the coding of the RNA molecules in the cells throughout the worm's body. Presumably, whenever the animal learns, the RNA is altered appropriately so that when regeneration takes place, *the altered RNA builds the memory into the regenerated animal right from the start.*" [Italics mine.]

A startling idea then occurred to McConnell. As he tells it: "In 1957 when we got our first results on retention of learning following regeneration, and came up with our chemical hypothesis, it seemed to us that we might be able to transfer a memory from a trained animal to an untrained animal, if we could somehow get the right chemicals out of the first worm and into the second. We spent several years trying to test this admittedly wild notion without much success. First we tried grafting the head of a trained animal onto the tail of an untrained planarian, but this never worked very well. If one reads introductory zoology texts, one often gets the notion that this little operation is most easy to perform. Sadly enough, the best average on record is three successes out of 150 attempts, and we simply did not have 150 trained worms to waste. We tried grinding the trained worms up and injecting the pieces into untrained animals, but we never could master the injection techniques. It was only some time after we began this work, that it occurred to us that we could let the animals do the transferring for us. For, under the proper conditions, one worm will eat another. And since planarians have but the most rudimentary of digestive tracts, there seemed an excellent chance that the tissue from the food worm would pass into the body of the cannibal relatively unchanged.

"So with Barbara Humphries as our chief experimenter, we

conditioned a number of worms, chopped them into small pieces and hand-fed them to untrained cannibals. We also took the precaution of feeding a number of untrained worms to untrained cannibals for a control or comparison group. Our first study gave us such unbelievable results that we immediately instituted several changes in our procedure and repeated the study, not once, but four times. And each time the results were quite significant —and still rather unbelievable. I should mention, before going any further, that the chief procedural change we made was the institution of a 'blind' running technique which guarded against experimenter bias. Under this blind procedure, the person actually training the worms never knows anything about the animals he runs—we follow an elaborate coding system in which each animal's code letter is changed daily. . . .

"I would also like to mention a couple of fortunate mistakes we made, which do not prove anything, but which are interesting in their own right. One time our elaborate coding system broke down and a control animal was fed a piece of conditioned worm. For several days prior to this feeding, the control animal had been responding at an average of two or three times out of any twenty-five trials. Immediately following the inadvertent meal of conditioned tissue, the animal performed at criterion level, giving twenty-three responses out of the next twenty-five trials."

One of the oldest religious activities of mankind, even before homo sapiens became recognizably established as such, was the act of ritual cannibalism.

As man developed and then became dependent on tools, he became a manipulator, of ideas as well as objects. This talent for the manipulation of abstractions underlies that quality of mind we call "reason." As man became ever more the reasoner, he also became ever more the manipulator, altering, through the exploitation of reason, the environment in which he lived, removing himself willfully from his ancient molecular memories, his intuitions concerning himself and his place in the cosmos. Reason is the enemy of intuition. But intuition persists. The intuition that persists even today, in the sophisticated Christian West, that there is a knowledge-substance or a memory-substance in the flesh, is testified to by the persistence of the ceremony of the Eucharist, in which a symbolical act of cannibalism is enacted as the wine and wafer are transformed into the flesh and blood of the Godhead. The intuition underlying the symbolic act of cannibalism is inseparable from the intuition inherent in the notion of resurrection—both involve a transformation of the flesh-as-knowledge.

The chromosome theory of inherited memory does no violence to this intuition, for every cell of the flesh contains chromosomes. Yet they are contained within a protective envelope, and in terms of a basic analogy bear the same relationship to the cytoplasm of the cell, as the skeleton does to the flesh of the body. It is the armature around which the structure assembles itself. Compared to the flesh and its relatively great potential for change, the skeleton is rigid and unchangeable. The flesh is far more responsive to the body's moods and circumstances.

Sonneborn's great contribution to our understanding of ourselves was his disclosure that, though the chromosomes still maintain the general structure of the cell in heredity, playing the role of armature, there is hereditable information or knowledge carried in the flesh or cytoplasm.

The Greek word *mermeros*, from which our word *memory* ultimately derives, originally meant "baneful, warlike, crafty, mischievous." Webster's New International Dictionary (2nd ed.) defines the word as "anxious," and associates it with the word *martyr*. Sager's work demonstrated that these ancient and active meanings of the word are more amenable to describing the molecular memory and its function than the modern, passive connotations connected with the word in English, with its implications of an inactive data-storage role. Sager demonstrated, in her bacteria of green algae, that the molecular memory plays a crafty, even perhaps a defensively warlike, role in adapting the organism to confront hostile circumstances. It would appear that the earliest, most intuitive meanings of the word were the most literally true, and that generations of reasonable, reasoning men, divorced from the participation of their intuition, have denied the essential reality of the meaning of *memory*.

The paradox of twentieth-century science consists of its *unreality* in terms of sense impressions. Dealing as it does in energy transformation and submicroscopic particles, it has become a kind of metaphysics practiced by a devoted priestly cult—totally as divorced from the commonsense notions of reality as was the metaphysics practiced by witch doctors and alchemists. It is not at all odd then, to discover that the closer we come via the scientific method to "truth," the closer we come to understanding the "truth" symbolized in myths and ritual practices concerning the habitation in the flesh of knowledge and understanding. The so-called body-mind problem does not occur to primitive peoples as being a problem. In the Orient as well, the dichotomy is not seen quite so distinctly as in those Western terms, which view the mind as being enclosed within the skull and all that lies below the neck as being merely an inefficient vehicle for transporting the mind from task to task.

Sonneborn and Sager proved the existence of memory-in-the-flesh at the lowest possible level of memory function, investigating the role of the molecular memory in the behavior of the single cell. McConnell, through his talent in exploiting a happy accident, managed to leapfrog across millions of years of evolutionary history, by demonstrating that even after eons of alteration the role of memory as a substance inherent in the flesh was relatively unchanged in principle in a multicelled organism equipped with a central nervous system.

McConnell showed that the RNA molecule might play a role in the memory retention existing in planarian tails. Hjylmar Hyden, the great Swedish biochemist, had earlier hit upon the suspicion that this same RNA molecular memory factor might be operative in humans. It was Hyden's work that stimulated Corning and John to soak their planarians in ribonuclease, thus destroying their RNA, in order to determine whether this would play any role in their retention of learning. According to John Gaito a colleague who has written a review of Hyden's work, "He [Hyden] believed that the nerve cell fulfills its function under a steady and rapidly changing production of proteins with RNA as activator governing the molecule. He hypothesized that memory involved a change in the sequence of bases in the RNA molecule; this change occurs when one or more bases are exchanged with the surrounding cytoplasmic materials. Hyden reported that individuals with certain psychic disorders have smaller amounts of RNA and protein in ganglion cells of the central nervous system than do normal individuals."

Also working with humans, not flatworms, two Canadian physicians Ewen Cameron and Leslie Solyom, using Hyden's hypothesis as a jumping-off point, wondered whether there was an RNA molecule basis for human memory. They found "that administration of RNA to individuals with pre-senile, arteriosclerotic, and senile syndromes (with some degree of memory impairment) brought about memory improvement. These changes involved almost total retention in some cases. When RNA was discontinued later, memory lapses occurred."[1]

Linus Pauling, the Nobel laureate, proposed in 1960, before the notion had become widely accepted, a general statement in layman's terms of what occurs continually in the human mind and body during every moment of life. "I believe," he writes, "that

[1] Controversy surrounds this experiment by Cameron and his associates. The RNA, which was administered orally, had been obtained from ordinary yeast. It is assumed that, in the course of its passage through the stomach and the gut, before reaching the bloodstream the RNA molecule was broken down into its component nucleotides. Then, if Cameron and his associates' data were correct, it must have been reassembled into a utilizable RNA molecule. According to critics of this experiment, no reasonable hypothesis is presented to account for the inability of the body to synthesize RNA directly from ordinary food sources. In other words, could the patients have been fed ordinary yeast instead of RNA derived from yeast? There has also been some criticism of the experimental techniques employed by Cameron and his associates. It is claimed that certain patients were arbitrarily selected, that there was no random sample, etc.

thinking, both conscious and unconscious, and short-term memory involve electro-magnetic phenomena in the brain interacting with the molecular (material) patterns of the long term memory, obtained from inheritance or experience." This handsome statement actually describes nothing except an epiphany, though the term *epiphany*—an apparition of God—rings hostile to the modern ear, invoking as it does the Gregorian tones of irrational demands upon credulity. But for life itself, for memory, to have emerged from out of the wasteland of degenerating matter—the raw fact that it did, strains credulity as well.

We remember, therefore we are.

PART TWO

The Population

[4]

The Species Problem

Darwin's awesome contribution to biology depends in large part on his understanding of the dimension of time as marking the stages of a process. Conceptually he broke through that membrane separating the past from the present. The past *is* in the present, but the present is not necessarily in the past. Not that Darwin was the first or the only man who ever recognized this truism—many years before, the Greek philosopher Heraclitus had remarked that we walk and do not walk into the same river twice. This ambiguous description of the flow of time was not congenial to Darwin. He was intent on demonstrating that we can never walk twice into the same river. This is a difficult idea to contemplate, for the river is *there* and the mind moves toward perceiving things as they appear to be, taking their appearance to be their reality. It often comes as a shock to look at the childhood photographs of friends we have come to know as adults. We leaf through the album of their childhood and have a sense of a vaguely familiar presence, one face among that mass assembled for the school photographer that causes our eyes to return to it. But the face is marked by a distant strangeness, and there is no certainty in the identification. The face of the child persists in the adult but the act of imagination required to make the association resembles a kind of backward prophecy.

This experience of ontogeny, of individual development, is common to all of us. Perhaps Heraclitus could have better said that we meet and do not meet the same man twice. We recognize this in ourselves, but are not always conscious that this process occurs along a broad front, and we are struck with consternation at class reunions.   We know that time has worked changes in us and our appearance, but there are invariably some of our old classmates whom we can recognize only by their names.   Far more accurately than any mirror, the changes in their faces inform us of what we have become.

When time is examined in this aspect, as the measure of change, it becomes an intangible that unhinges the mind wishing to examine it.   The process being enacted in time can only be perceived comparatively.   A watched pot never boils, the hands of a clock do not move.   We notice the growth of our children only as their clothes shrink.   Ontogenetic time can only be seen retrospectively, by measuring one stage against another, later one.

But if ontogenetic development is difficult to observe, phylogenetic development, the changes wrought in a population by time over the course of generations, is impossible to see directly.   The mind can only grasp it as an idea, an abstraction.   We know that these changes occur in the same way that we know the earth is a sphere and revolves around the sun.   It is an abstraction, there is no simple testimony from the unaided senses which will either confirm or deny it.   If we cannot imagine the pace of those moment-to-moment cellular changes taking place in us as we grow and age, then how can we possibly imagine those infinitely minute alterations which occur as one generation succeeds the next?   As our scale of observation of these changes expands in time, we must include ever larger reaches of forms, for the only meaningful lineages are those which include entire populations. Can one discover the size and shape and color of a beach from a single grain of sand—or even a handful of grains?

The smallest developmental entity which can usefully be employed as a unit for this kind of measurement becomes the species.   This was Darwin's problem and he bequeathed it to us, for the problem remains as acutely problematical today as it was for Darwin over one hundred years ago.   To the layman's first glance, it might seem simple enough; the word *species*, after all, is simply a word.   But it is a collective noun, like the word *forest*, a word to which children respond far more directly than their elders.   To children the word forest is dark and frightening; they love to hear stories of people lost in forests, because such stories represent the primal anguish of their lives.   In a forest the individual trees lose their particular uniqueness, becoming identical through the press of their numbers.   The symbol of the forest represents the trivia (in the original meaning of the word— a *Y*-shaped crossroads) of the child's life, of all our lives.   The main stem of our own existence is impelled forward along two alternative routes: the one leading toward isolation and the compulsion to exploit the particularities and uniquenesses which represent the essence of us as individuals; the other driving us to become part of a communion—a collective larger than the self— to lose our sense of separateness in the flock, the herd, the group, the tribe, and so on.

Because this double goal is such a trivial commonplace of life— requiring a subliminal decision every time we open or close the door to another room, or every time we meet another's glance or avert from it—we tend to overlook that all animals from the most minute microbe to the largest blue whale must confront the identical dilemma—as individuals and as populations.   Nowhere is this dilemma more directly perceived than in the struggle to find an acceptable definition for that simple word *species*.

Prior to Darwin's day, judgments regarding species classifications were based on physical form and appearance.   It was well known that forms varied with geography; there were tropic forms, and arctic forms, and so on.   It was Darwin's great insight, while voyaging in the Pacific aboard H.M.S. *Beagle*, to discover that time and space worked together to produce the diversity of living forms.   Just as the individual human requires privacy in order to exploit his inner uniqueness, so the species requires the privacy of isolation in order to become "diverse" from other similar populations of animals existing elsewhere.

As usual it was at first the obvious, overt form of isolation that attracted the attention of evolutionary theorists—the geographical, or spatial isolation, the kind of isolation provided by various natural barriers, oceans, mountain ranges, deserts, etc. Only recently has the attention of zoologists been drawn to the other, more covert forms of isolation—behavioral isolation. In his book *Animal Species and Evolution* published in 1963, and generally accepted as the most complete and exhaustive review of modern evolutionary theory, Ernst Mayr flatly states: "If we were to place the various isolating mechanisms of animals according to their importance, we would have to place behavioral isolation far ahead of all the others. . . . Ethological barriers are the most important isolating mechanisms in animals. . . . A shift into a new niche or adaptive zone is, almost without exception, initiated by a change in behavior. The other adaptations to the new niche, particularly the structural ones, are acquired secondarily. . . . Most recent shifts into new ecological niches are, at first, unaccompanied by structural modifications. Where a new habit develops, structural reinforcements follow sooner or later."

If modern zoology has now, at long last, come to consider behavioral isolation to be the principal operative factor in the development of animal species, how much more certain may we be that it was this factor, the exclusion of all others, which determined the speciation of that odd group of hominoid primates from within the larger population of terrestrial apes.

What was this "new habit"? Was it a new, more comprehensive communication system—speech which permitted the evolution of highly complex, yet flexible forms of social organization? Was it that new and special sense of time and causal relations existing within time sequences, which permitted this population to comprehend the principle of the tool? Many animals are known to use natural objects as tools, but only hominoid primates were possessed of sufficient foresight to manufacture their elaborate permanent tools far in advance of need and carry them about in anticipation of possible use. This talent for projecting causal sequences in the future is unique to human beings, and without it tools could not be.

How the sequence occurred—whether the manipulation of objects as tools preceded the manipulation of ideas into the symbols of speech, or whether the sequence was reversed—does not really matter much. However it happened, it happened more or less concurrently and served the important function of effectively isolating this new population of hominoid primates from all other similar populations.

There is good reason to believe that the first tools man employed were not primarily used as weapons. Judging from the daily routines of such terrestrial primates as the baboon (and the partially terrestrial chimpanzee) as compared with those of existing hunter-gatherer peoples still living a Stone-Age life style in South Africa, anthropologists now suspect that man's first tool was some kind of digging stick which enabled him more easily to unearth those edible roots and tubers that served as the mainstay of his diet. But it requires only an alteration of intention to transform such an agricultural instrument as a pitchfork (the modern version of a digging stick) into a deadly offensive weapon.

Throughout his long and tormented history, man has puzzled more over the underlying causes of that strange collective behavior known as war than he has about any other of his collective activities. It is difficult for humans, lacking the natural armament of predators—canine teeth and talons—to fight wars with their bare hands. One cannot imagine wars without imagining weapons, and we have come to make that connection causal, and to hold disarmament conventions in the hope that, by destroying weapons, we shall destroy war.

During the nineteenth century the notion that wars were precipitated by economic greed was popular. In this century the prevailing notion is that ideological conflicts provide the basic rationale for war. There is no doubt that all these factors play their part as vectors causing entire societies to drift into that destructive collecting enterprise. War is like a boat drifting on the wind and tide.   But before the boat can be said to drift, it must exist.

It is this for which we must search—the existence of this basic buried tendency—and it is here that the animal evidence may be useful in creating for us conceptual models of how the particularly human behavior of war may have become elaborated, over the course of millennia, into that ritualized pattern it now displays.

It is among the social insects such as bees and ants that we find the closest parallels.   There are, among certain species of ants, "soldier castes," endowed with special structural refinements of their bodies that enable them to perform the unique social role of making war.

Among ants the evolutionary response to the principle of collective aggression was the emergence of the soldier caste. Among human beings the response to the principle was behavioral.   What is the principle?

The principle lies sleeping in the very essential core of the species problem.   Stated in grossly oversimplified terms the problem of the species and the process of speciation has to do with the reconciliation of two opposing drives: one impels the individual and the population of similar individuals toward isolation, toward privacy and separation; the other, its opposite, impels the individual to become part of a larger aggregation, and to surrender uniqueness within this communion, acquiring the characteristics of conformity to this larger grouping.   As these opposing drives have become expressed in behavior and in changing form appropriate to changing behavior, that phenomenon known as *speciation* occurs.

For Carolus Linneaus, that titanic Swede without whose work the modern theory of evolution could not have come into being, the species problem was strictly formal.   Trained as a botanist, he was intellectually predisposed to overlook behavior.   As Loren Eiseley describes him, he "had a poetic hunger of the mind to experience personally every leaf, flower, and bird that could be encompassed in a single life . . . a new Adam in the world's great garden, drunk with the utter wonder of creation." His botanist's mind, however, was fixed on form. For Linneaus dead things took on a new life, just like the puppet Pinocchio with a new name. From his time until just recently naturalists brought back to their museums from the far corners of the earth the bones and skins of dead animals, which were placed into species categories on the basis of their appearance alone.

Offhand, this might seem eminently sensible. Obviously, penguins differ from peacocks because they inhabit different geographical locations with differing climatic conditions. The differing problems posed by their differing environments were solved differently for each species of bird, with the result that eventually their appearance came to differ.

But what about those creatures that inhabit the same locale? There is that animal, for example, known in the King James Version of the Bible as the cony, and more formally as the Hyrax, which inhabits the Near Eastern and the African savannas. The technical name *Hyrax* derives from the Greek *hyrakos* meaning mouse; the English *cony* derives from the Latin *cuniculus* meaning rabbit. Insofar as its appearance goes, either name adequately describes the animals exterior character. It looks obviously like a rodent—either a large, furry, woodchuck-size mouse, or else a nonjumping, short-eared rabbit. However, because of the interior details of its structure, it is unmistakably an ungulate, not a rodent; its closest living relatives are the rhinoceros and the elephant. And moreover, both historically and at the present time, it has shared its habitat with the rhinoceros and the elephant.

This study of differences and relationships with the ultimate aim of cataloguing the diversity of life is known as the science of systematics, or taxonomy. It is an "unnatural" act in that it is an imposition on nature; it is a function of the human mind—that part of the mind, at least, that responds to language. Language is principally composed of nouns and verbs and one must find names for things in order to communicate information about them. But any attempt to substitute a linguistic experience for the real event of a happening is doomed to failure. How can one describe the color green to a man born blind?

Nature does not classify. All the colors run together in a rainbow, and all life runs together with no sharply demarcated boundaries anywhere. Even the borderline between life and nonlife is blurred. Are viruses alive or inert? The borderline between the two great kingdoms of life is also blurred. Are the slime molds animal or vegetable?

In such instances, where the flowing borders between the forms makes morphological distinction all but impossible, the taxonomist must make his judgments on the basis of behavior. Viruses are living forms because they possess the molecular memory and are capable of mutation. Slime molds are animal because they are volitionally mobile, moving to and fro under their own power.

But just as in the rainbow, where in the center of the red stripe the red is unmistakably red, and in the yellow, unmistakably yellow, so the majority of animals placed within the larger categories of kingdom, order, class, and genus, can be defined on the basis of structure, appearance, and general behavior. To make classification judgments concerning such obvious disparates as the cony, the rhinoceros, and the elephant is not difficult. What difficulties there are in such a classification problem concern the discernment of relationships. For the point of species separation occurred so long ago in linear time, and the histories of their mutual divergence have been so exhaustively extensive, the relationship has become vestigial as is that between ourselves as humans and the squirrel-like tree shrew, both of whom are classified within the primate order.

It is when one descends from the larger categories into that of species distinctions that the neat, formal lines of description that so delighted Linnaeus vanish beneath a smear of uncertainty, a snarl of doubts caused by our comparative ignorance of the process of divergence, the infinitely small-scale mechanics which cause one population of animals to suddenly separate its mutual history from that of the larger population and strike off on a new historical adventure of its own.

Ernst Mayr cites the parable of the thrush family of songbirds in New England. There are four separate species of *Catharus* inhabiting the woodlands and meadows of the northeastern states; the veery, the hermit thrush, the olive-backed (or Swainson's) thrush, and the grey-cheeked thrush. Mayr writes: "those four species are sufficiently similar visually, that they confuse not only the human observer, but also silent males of other species. The species-specific call notes, however, permit easy species discrimination. . . ." They differ, these four species of birds, not nearly so much in their appearance as in their behavior—in their songs, their foraging and migratory habits, and in their preferred areas for nesting. The veery breeds in bottomland woods with lush undergrowth, while the grey-cheeked breeds in stunted northern fir and spruce forests. The hermit forages for food on the ground at the edge of the forest; the olive-backed in the trees of the interior forest. The veery and the hermit thrush build nests on the ground; the olive and the grey-cheeked, in trees. The Cornell University ornithologist William Dilger has discovered a number of minute anatomical differences; the various species are beginning to acquire differing lengths of leg and wing; the relative shapes of their beaks differ. But most importantly, as Mayr says: "No hybrids or intermediate among these four species have ever been found. Each is a separate genetic, behavioral, and ecological system, separated from the others by a complete biological discontinuity, a gap."

Within the flowing continuity of nature one is always aware of empty spaces. There is empty space between the grains of sand that make up the continuity of a beach, there are gaps between the blades of grass on a lawn, and there are gaps between populations of animals. It is just this sort of gap, this discontinuity, this strange, sudden appearance of an emptiness between two populations which attracts the interest of modern zoologists.

This gap must appear before the two populations can become diverse from one another. It is just the appearance and then widening of gaps such as this, which, according to the modern, so-called synthetic theory of evolution, has created those differences between life forms which we see around us, those differences for example, between the cony and the elephant.

At one time our ancestors were members of a primate order which must have comprised one single population of animals that looked very much like modern squirrels, though the main fixture of their diets was probably insects, not nuts and berries. Gradually, as their *umwelts* expanded and their curiosity grew, certain groups of these animals began exploiting new aspects of their environment. Their forms then changed in response to the new needs of their lives.

Today it is taken for granted that evolution occurs; the idea of phylogenetic development is accepted as is the idea of ontogenetic development. But our understanding of the time scales involved is completely missing. We simply have no idea how long it takes for almost anything to happen in evolution—or why rates should be so uneven. It would seem that certain populations of animals are halted in a hiatus of attentive expectancy for very long periods of time. In our order of primates, for example, there are the families of tree shrews, lemurs, lorises, tarsiers; for the most part squirrel-like insectivores which have not changed nearly so radically from their fossil forebears as have we. Darwin's hypothesis, coupled with the discoveries of modern genetics, have given us a retrospective comprehension of what must have happened. We know that branches grow apart from the trunks of trees, and that their individual growth will depend in part on the vicissitudes of their particular location; they will either flourish if they have access to light and space, or remain stunted if they are deprived. The analogy persists in the "family trees" of animal relationships, and insofar as evolutionary theory goes, the most fascinating part of the process is the one that occurs right at the point of branching.

How does it occur?   Geographic isolation is not the answer. "San Francisco Bay," Ernst Mayr writes, "which keeps the prisoners of Alcatraz isolated from the other [human] inhabitants of California, is not an isolating mechanism, nor is a mountain range or a stream that separates two populations that are otherwise able to interbreed."

The species problem, when finally examined here, at its roots involves an understanding of sexual behavior, of compatibility and incompatibility.   What happens is that suddenly a splinter portion of any given population no longer chooses to interbreed with the main body.   The members of this splinter party suddenly begin to interbreed exclusively with one another, rejecting potential mates from outside the group.

As a result they share the genetic memory of their communal experiences with one another and develop their particularities aloof from the parent population.   In the past fifty years the fallacies inherent in the Linnean system of classification on the basis of appearance have caused zoologists to formulate a classification based on something other than appearance.   Appearance can vary with the individual.   For example, prior to 1950 the weasels of North America were classified into twenty-two different species.   A patient zoologist, Eugene R. Hall, after observing them for a long period, in 1951 published a 466-page paper which finally convinced his fellow taxonomists that the weasels of North America really belonged to only four separate species.   There appeared the typical species gap between four weasel populations.   All the rest of the varying animals were merely subspecies, or races.

How then is the species defined if not on the basis of appearance?   How is this gap between populations perceived? It is perceived in sexual terms.   Ernst Mayr defines the species as follows: "The species, finally, is a genetic unit consisting of a large interconnecting gene pool."   It is the word *interconnecting* which is the operative word in this definition.   So long as the pool is interconnecting, the population is capable of fermenting—

producing its own interior variations.   Only when it ceases to be interconnecting, when a discontinuous splinter becomes a separate fragment, does it become a species embarked on the road toward extraordinary differentiation—such differentiation for example as exists between the gibbon, the gorilla, and ourselves.

To the layman, however, these classification games may seem to be sterile exercises without any truly functional uses.   Since it is merely a matter of names, what difference does it really make if the red fox differs from the grey fox on the basis of a species category, or merely a racial category?   The functional uses of species categories can occasionally come into play when the animals involved are important to man.   One such instance took place in the 1930's during an exhaustive study of malaria distribution in Europe.

No less an authority than Sir William Osler, the famous British physician and medical historian, has called malaria "the single greatest destroyer of mankind," and considering the vast array of ills that human flesh is heir to, including the malice of other humans, the odd little parasite which produces the disease has accomplished a remarkable record.   It may yet prove to have been the single greatest destroyer of vertebrates generally, and therefore one of the most potent pruning hooks of the natural selection process, for contrary to popular notion, it is neither restricted to the tropic zones nor to warm-blooded mammals. Fish, reptiles (particularly lizards), and birds have all been discovered suffering from the disease.   It has spread so far from the tropics, which may well have been its original birthplace, that penguins living well below the Antarctic circle have been discovered suffering from the disease. Though in this latter instance, instead of a mosquito vector being the intermediate host, a louse or mite is suspected of carrying the parasite.

Several authorities believe it has had an incalculable effect on human history, particularly the development of modern civilizations in the temperate zones where there is a shorter summer

activity season for the adult female mosquito. The disease is
also considered partially responsible for the decline of several
great Mediterranean cultures, particularly that of Athens, between
the fifth and third centuries B.C., where the literature abounds
with excellent clinical descriptions of the classic symptoms. It
may have worked analogously in other animal phyla, sapping its
victims of energy, creating brain damage with resulting sluggish
neurological reactions, so that many populations of animals which
were superbly fitted to cope with other elements of their environment may have perished because of their inability to withstand
the ravages of this disease. Paul F. Russell, Chairman of the
World Health Organization's Committee on Malaria, estimates
that the total number of contemporary cases of human malaria, as
recently as 1952, ran about 350 million, or roughly 6.3 per cent
of the world's population.

Causing the disease is a single-celled protozoa—an animal of
the genus *Plasmodium*, in which there have been described some
fifty species. This is a *de facto* classification based on such considerations as the choice of vertebrate host, the descriptions of
symptoms (particularly their periodicity) and the locale of the
victim. Whether these classifications represent true species
categories, or subspecies (racial) categories, or whether they
represent differing aspects of the same general population is still
unknown. If judgments were to be made on the basis of
appearance alone, there would be only four major species; and
three of these were identified as far back as 1885 by the great
Italian pathologist Camillo Golgi; little progress has been made
since then.

The principal obstruction to certain species identification of the
parasite is its capacity for masquerade. It is *cyclomorphotic*; there
are distinct alterations of form, appearance, and behavior from
one generation to the next, and there is good evidence for believing these alterations are triggered by environmental factors (in
roughly an analogous way to the activation of certain genes in the
varying hare, which produces a white coat in winter and a brown

in summer, depending on the length of daylight hours). It is therefore conceivable that the same animal might infect humans and monkeys, appearing and behaving differently in each host.

In its vertebrate host, the animal produces the major portion of its population inside the red blood corpuscles. There is a little-understood transitional phase when the animal enters the host and resides in one of the organs (the spleen and liver are favored locations in primates), and there may be some reproduction—enough to give the population a start—somewhere other than in the red blood cell. But this latter location is where the vast bulk of the population resides. The red corpuscles are living cells shaped like a concave disk, composed mainly of hemoglobin. They are produced by specialized cells within the marrow of the long bones of the body. The corpuscles live for about fifty to seventy days, and upon dying, their corpses are destroyed by processes occurring in the liver, the spleen, and the lymph nodes. These latter organs are the first to suffer from the effects of the disease. They become enlarged and overworked in an attempt to cope with the increased mortality of corpuscles, and if the parasite population is not stabilized by one means or another, they ultimately fail in their functions. But before this happens, other unpleasant effects are felt. Hemoglobin, the stuff on which the parasite lives, is an oxygen transport material transferring the oxygen taken in by the lungs to the various body tissues requiring it. If a person dies of malaria, his body literally smothers to death. The capillaries become distended, clogged with dead and dying cells, blood flow is impeded, and the brain—one of the most voracious users of oxygen—will be damaged by oxgen deprivation. If the damage is not too extensive, the brain can transfer functions from destroyed tissues to intact ones, but in many cases of malaria fatality, the mortal blow to the individual was dealt in the brain itself.

The parasite enters into its vertebrate host as a tiny hairlike creature, called a sporozoite, which matured within the stomach of a female mosquito. The sporozoite is injected into the vertebrate as the mosquito spits its irritating saliva into the wound created by its proboscis mouth. In the late nineteenth century Alphonse Laveran, a French medical officer stationed in a military hospital in Algeria, first discovered the creature. Considering the tremendous technological advances which have occurred since then, surprisingly little more is known about the parasite. Laveran's paper dated December 15, 1880, describes the following: "On the 20th of October last, while examining by microscope the blood of a patient suffering from malarial fever I observed in the midst of the red corpuscles the presence of elements which appeared to be of parasitic origin. . . . I have described the elements as bodies . . . elongated elements, more or less tapering at the extremities, often curved into crescents, sometimes of oval form. They measure 8–9 thousandths of a millimetre in diameter. They are colorless except towards the central part, where there exists a blackish spot formed of a series of rounded granulations which appear to be pigment granules. Some of the bodies show on the concave side, a pale curved line which appears to tie together the extremities of the crescent."

The crescent bodies which Laveran described were the asexually reproducing forms of the protozoa now known as schizonts. This schizont form is the one that the animal assumes to best exploit its vertebrate host. As it grows within the red blood cell, it produces forms called pseudopodia—literally *false feet*—which grow out in seemingly haphazard fashion from any part of the original schizont until the cell is completely filled with this Medusa-head mesh of living matter. These false feet now break away from one another; and as this happens they are given another name, *merozoites*. They become active as a swarm of snakes, and the blood cell ruptures under the pressure of their writhing. This cycle may take anywhere from one to three or four days—the period required is one of the species criteria. As the corpuscle ruptures, the merozoites escape into the surrounding plasma and each one actively squirms its way into another cell. Sometimes, if the population is dense, two or three may enter a single cell, but in that case future growth is stunted. The number of these merozoites which derive from any given schizont is also a species criterion. In the most virulent human form of the disease, the production averages about sixteen merozoitcs per schizont.

Almost the entire population of parasites reproduces simultaneously; this causes the classic symptoms of malaria—a sudden chill which leaves the victim blue-lipped and shaking with an ague which can last a few minutes or an hour, and which is followed immediately by a furious rise in temperature, up to 107°F, often accompanied by delirium, frightful aches in the bones, vomiting, etc. This paroxysm ends after about four hours, and is followed by a relatively tranquil sweating stage during which the patient may fall into an exhausted sleep. The victim may then be relatively without symptoms until the next period of schizont reproduction, which occurs between two or three or four days hence, depending on the species of parasite. The majority of these paroxysms occur, at least in the tropic forms of malaria, after midnight and before dawn. The reproductive cycle of the parasites is tuned to the favored time for the mosquitoes to seek out food. In temperate zones, or locations where the mosquitoes feed in the daytime or early evening, the symptoms of the disease occur then.

For some reason that we still do not understand, a few members of this parasite population do not assume the asexual schizont form. These are the "oval forms" described by Laveran, and it is in their interest that the cycle of the schizonts occurs when it does, during the mosquito's favored feeding time, for if these oval forms remain in the blood of a vertebrate they will perish. They are sexual animals, and in order to complete their mating activities, they require a special setting, the unique and particular environment that obtains within the gut of a female mosquito. Not any mosquito—but a particular species of mosquito—and it was this host preference of the malaria parasite that led obtuse human entomologists to separate into species categories several populations of anopheles mosquitoes that had appeared identical.

The particular set of conditions which prevail within the mosquito gut have never been reproduced in the laboratory, so no one has any idea what these requirements may be. But the full cycle of parasite mating activities can only occur in this setting. Some of the preliminary transformations of these oval forms into free-living sexual animals can occur outside the mosquito's gut— on a microscope slide—though it is not believed that they can occur in the bloodstream of the vertebrate host. The transformation happens fairly quickly, within a span of ten minutes or so. Some of the oval bodies contain male spermatazoa; they burst, rupturing the corpuscle as they do, and releasing a swarm of swallow-tailed sperm into the surrounding plasma. These active creatures, which swim about lashing their double tails, are known as microgametes. The other oval bodies are females. Under the microscope, if one is careful about the staining procedures, one can differentiate between the two types of oval bodies. One can sometimes see within the oval envelope which surrounds the microgametes the compressed community of males, swarming and writhing, at least at a certain stage of maturity shortly before the males are released.

The female oval bodies, known as macrogametocytes, appear dense and more coherent; they also grow larger and finally explode out from the cell which surrounds them through the sheer increase in bulk. Once released from the blood cell they are mobile, capable of some movement; given a surface to cling to, they screw themselves about with a hideous scrunching worm-like movement. In the mosquito gut, and occasionally on a microscope slide, they will develop still further, growing an odd humping protuberance, which seems to exert an attraction for the males. If any microgametocyte finds himself in the vicinity of this hump, he lashes himself toward it and enters it like a battering ram. Then genetic materials are exchanged and the female is fertilized. As this happens, she becomes endowed by scientists with still another name, oocyte, and from this point on any further development must occur within the gut of an appropriate species of mosquito.

As the mosquito sucks up its meal of blood, it takes into its gut, along with the sexual forms of plasmodium, the asexual forms, schizonts and any free-swimming merozoites which happen to be about. These creatures are digested along with the ingested vertebrate's blood. Only the sexual forms resist digestion.

After being fertilized, the female screws herself deeply into the mosquito's gut wall where, if conditions are proper, she will be enclosed within a cyst. She now grows to a huge size, large enough to be visible to the naked eye. She mushrooms through the mosquito's gut wall, and continues growing from the other, outer, side. Dissecting a mosquito to remove the gut is not as difficult a procedure as may be imagined. It can be done by anyone who knows the trick with two ordinary pins on any smooth surface. The gut comes right out like a tiny piece of brown spaghetti, and the mature oocytes can clearly be seen protruding like mushrooms on small stalks from its outer wall. Eventually these oocytes burst, releasing a horde of small hairlike animals similar in appearance to microgametes. But their lashing tails, instead of extending from the rear of the body, extend fore and aft; the creature looks like a transparent snake with its opaque head carried amidships. The sporozoites now swim actively around within the fluids inside the mosquito's body cavities until eventually some of them enter the salivary glands. Once there they seem content to remain even though, after a period of time, the population may become quite dense. There they collect and wait for the mosquito to spit them out along with its irritating saliva into the body of a vertebrate animal.

It is believed by some malariologists that sporozoites have the ability to penetrate tissue, finding their way into the nearest capillary by burrowing right through the flesh. It is conceivably possible for a person to contract malaria even if he is not bitten by a mosquito—even if he squashes the insect on his skin before it has a chance to bite, the sporozoites may still be able to enter into his bloodstream and infect him. Once in the bloodstream the parasites migrate rapidly to the liver or the spleen (in the case of primates), where they begin reproducing and eventually the population spills out into the red blood cells where the parasites travel throughout the body as a whole, each enclosed in its hospitable capsule, a red blood cell.

Interesting as this recital of the malaria parasite sequence may be, both in itself and as an example of parasitic adaptation, it may still seem far removed from the species problem. Yet it was largely through the concerted efforts of several malariologists that the modern "sexual" concept of the species, the so-called New Systematics, developed. Mounting field trips into remote areas with special equipment to study animal sexual relationships is an expensive business. While involved in their mating activities, many animals are peculiarly vulnerable to predators; they make strenuous efforts to accomplish the mating act under conditions of great privacy—in darkness, in inaccessible locations, etc.—and at this time more than any other, they resist observation.

But since detailed and intimate knowledge of the entire generative cycle of both the mosquito and its infectious parasitic passenger was of such crucial importance to the health of mankind, no expense was spared in sending men and equipment anywhere in the world where the mosquito was suspected to exist so that all the information possible about the mosquito's habits generally, its round of daily activities—its feeding habits, resting habits, those habits connected with copulation and oviposition—could be obtained for the purpose of destroying the mosquito more effectively and economically; and with their destruction, hopefully, control of the disease they carry.

As with most blood-sucking insects, it is the female who requires the blood for the maturation of her eggs. The male anopheles drinks the juices of plants and fruits. In the laboratory it feeds happily on apples and raisins; the skins of these fruits represent the limit of the penetrating powers of its proboscis, which is blunter and more flaccid than the female's.

The female proboscis is a complicated instrument composed of comparatively rigid members—the jaws, which have grown together into a tube during the course of evolution. The lips remain flexible members; the insect uses them as a retractable guide-sleeve while introducing the proboscis into tissue—for the act of biting is an introduction, an insinuation; it is not a stab. The tip of the proboscis, for the final third of its length, is more flexible than the rest, something like the tip of a fishing rod, and once the main entrance into the tissue has been made, this tip searches for a capillary at an angle of about 45°, first probing in this direction, then being withdrawn and redirected into another direction, and so on, until it strikes a capillary. Then the saliva which contains some decongestant properties is injected to "thin out" the blood and the insect begins sucking it up, often taking blood in the amount of its own body weight. After this it flies away to a resting place where it shall remain for at least two days (depending on species) until the blood is digested and the eggs matured. The next act is that of oviposition, and for this the insect must fly to a body of water (a rain-filled hoofprint is enough) to lay its eggs, for the larva which shall emerge from the egg is an aquatic creature.

It has been reckoned that the lifespan of the average mature female anopheles is somewhere around a week or so. One laboratory specimen has been maintained in the lab for eighty-six days, during which time it had six blood meals and laid six clutches of eggs; but this is considered exceptional by malariologists. The general feeling is that only the rare mosquito has more than two blood meals during the course of her lifetime. Here is another example of how absurdly unfavorable the environment of the plasmodium parasite would seem to be: the time required for the development of the oocysts in the gut of the mosquito varies with environmental temperature, shortening as the temperature rises, but on the average it seems to take about a week. The average parasite has then only *one* opportunity to re-establish its population within a new vertebrate host. And yet the prevalence of the disease testifies to the operational effectiveness of this seemingly unforgiving system. One chance for success is all that most animals get, and this one chance is sufficient.

The species problem only entered into malaria studies as the carrier of Europe, *Anopheles maculipennis* began to be intensively studied during the 1920's and 1930's. *A. maculipennis* is unmistakable, a small, rather darkly speckled insect sitting high on its spindly legs and pitched at a distinctly downward angle. Its rearmost legs are normally held up in the air off the surface, even when the insect is resting, not preparing to bite, and the body is tilted downward as though it intended to stand on its head. Most resting mosquitoes carry their body horizontal to the surface. As malariologists all over Europe concentrated their attention on *A. maculipennis*, certain inconsistencies became apparent. The first disturbing note appeared in 1920 when the French malariologist Emile Rouboud published a paper noting the abundance of *A. maculipennis* in many parts of France where malaria had never been reported. Quite independently, in the next year, Carl Wesenberg-Lund reported the same situation in Denmark, as did Battista Grassi for Italy. Wesenberg-Lund believed that his Danish mosquitoes had changed their food preferences. Rouboud and Grassi were closer to the truth in their speculations; they believed they had discovered a new "race" or subspecies of *maculipennis*. They believed they were dealing with a "race" rather than a new species, because there was nothing in the appearance of the adult to differentiate these benign insects from the disease carriers.

But then a Dutch entomologist, Nicholas H. Swellengrebel, given financial support for his studies as the result of an outbreak of malaria in the Netherlands, found that there was a difference between the carriers and the benigns. One could not discern it by comparing any two individuals from either population, but statistically he was able to determine average differences in the length of wing in each population; he named one of them, the suspected malaria carriers, "shortwings" (which has since been made over into formal Latin as *atroparvus*), and the other, the benign population, "longwings." Assisted by a colleague, Abraham de Buck, he went on, in the manner of a good ethologist, to discover that between the populations the entire repertoire of behaviors differed: their feeding habits, adult mating habits, their larval breeding places—everything differed except their appearance. Linneas' doctrinal shroud still blinded both these men, and they hesitated (they wrote that they considered it "inadvisable") to give separate Latin species names to these two populations. The next step was obvious: would they interbreed and would the offspring be fertile?

The shortwings would and did. Males would buzz any resting female within a cage no matter how small it was, and mount her. Matings between female longwings and male shortwings produced infertile offspring like horse and donkey matings. This was certainly diagnostic of a species difference. But the longwing males could not be induced to mate in small cages, or in large outdoor cages. They simply would not mate in captivity. This was curious. Were the pursuit of this curious phenomenon to be conducted purely for purposes of enlightenment, we would probably still be waiting for the answer. But a dread disease had struck a civilized nation, and so investigations leading to the solution of this problem were supported by serious men sitting on the boards of reputable institutions; it was no longer an eccentric obsession on the part of a handful of entomologists.

And so an American zoologist, Marston Bates, was dispatched by the Rockefeller Foundation to Tirana, Albania, where all the varieties of anopheles mosquitoes in Europe could be found in close proximity.

Bates was well aware that the entomological literature was filled with accounts of male mosquito swarms. Males would swarm together at certain times in an air-borne mass, sometimes so thick that because of one species, which tends to form like a cloud of smoke over houses, there were accounts of fire alarms being raised. Most of the time, however, the swarm is not quite so large or so dense, being roughly circular, about the size of a beachball and containing less than a thousand males. It was supposed that this swarming was part of a courtship ritual, and that it was a required prerequisite to the mating act; in some way, it served as an attractant to females. Bates was particularly struck by one report which stressed the particularity of swarming locations which, according to this observer, varied with species. Harrison Gray Dyar, the man whose report stuck in Bates's mind, was a trained entomologist with several new mosquito species carrying his Latinized name. The mosquitoes mentioned were not anophelline, but Bates felt the principle might well apply to anopheles. Dyar wrote that he had been living in the town of White Horse, in the Canadian Yukon territory. "The swarming habits of the common males at White Horse were constant and interesting. The town," he wrote, "is in the sandy level river-flat with a high bluff behind, formerly the river margin. On walking toward the bluff any still evening, males were encountered, first the *callithotrys* in the tops of small pines; next *prodotes* over open spaces between pines and willows; then, on reaching the high spruce trees, *lazarensis*, high up opposite the ends of projecting branches; and lastly in openings between tall spruce, over willow bushes, *punctor*, and an occasional *excrucians*, high up and flying wildly. At Dawson, *pullatus* appeared over willow bushes on the hillside any time after 4 P.M. that the sun went behind a cloud."

Bates settled himself in Tirana and built an insectary, a hut filled with wire mesh cages of various sizes, as small as a matchbox and as large as an orange crate. Outside, abutting the insectary, he built a large screened enclosure thirty feet long, twenty feet high, and fifteen feet wide. He writes that he got his first clue to at least one of the conditions that controlled the formation of swarms "when we went into the insectary in Albania one night to see what was going on. We switched on the light, which happened to have a dim bulb and a minute or so later we were surprised to find that a large swarm of *superpictus* [a malignant anophelline species] had formed in the middle of the room. We at once concluded that control of light was the answer to the problem of mating under laboratory conditions, and thereafter most of our attention was given to this factor.

The mosquitoes had to reach a suitable state of excitation, however, before a swarm would form, even if light conditions were optimal. These excitation states appeared to be a function of circadian rhythms. In some species the excitation state appears every three days; in between these three-day periods the males appear quite independent of one another, not all of them active at once—a necessary precondition for swarming. In other species the excitation state seemed unconnected to a circadian rhythm; swarming could be induced at any time if light conditions were right. "A change from bright light to dim light, or from darkness to dim light was equally effective," writes Bates. "Swarms seemed never to occur in complete darkness or in light of more than about ten foot-candles in a closed room, though under outside conditions, swarms were observed when the light was between ten and fifteen foot-candles. Once a swarm had formed, it was relatively stable and the light intensity could slowly be increased up to about fifty foot-candles before the swarm would disperse. After dispersal, however, the swarm would not reform until the light had again been reduced to about three foot-candles. Swarms dispersed at once if all lights were turned off, and would reform in thirty seconds to a minute when a light of about one foot-candle was again turned on."

As Bates transferred his attention to outdoor swarming in the larger cage, he found that the animals were even more delicately attuned to the cues offered by natural sunlight. A swarm of males from one population would form over the same spot on the floor over which another population of males had been swarming only minutes before, and had just dispersed. When swarming took place in the morning, the order of precedence was reversed (as was the progression from light to dark).

Try as he would, Bates was unable to discover why certain populations appeared to prefer forming their swarms regularly at certain locations within the cage.   Another observer, Dr. J. S. Kennedy, also working at that time in Albania, told Bates of having seen "a swarm of several hundred males inside a stable in Tirana, thirty to sixty centimeters above the floor in the lightest part of the room.   The swarm was complex, that is, with several different foci.   The position of the foci was apparently determined by patches of fresh dung.   In the second evening the swarm was in a different position in the stable and again over fresh dung."

After hearing this story, Bates conducted experiments hoping that scent signals might determine swarm locations, but he had no success in the cage.   Bates was very conscious of the fact that the condition of confinement might dramatically alter behavior so he leaves the possibility open that in the wild state scent may very well have a determining effect on swarm locations.   In the cage, however, it had none.   He did discover one peculiar trait in one of his populations: "We once observed that a small swarm had formed over a piece of white paper in a cage and we discovered that if this paper were moved slowly around in the cage, the swarm would follow.   In this case it was quite clear that the orientation was to light reflected from the paper, since this same result could be secured with a mirror, but not with dark colored paper."

He discovered that still other populations of males were attracted to vertical objects.   "The repeated observations of *Culex pipiens* over church steeples and chimney is an example . . . [one scientist] reports seeing this species swarming 'over low bushes,' 'on the lee side of a hedge,' 'near the gable of a building,' 'above a chimney pot,' and in a variety of other situations."   In his outdoor cage Bates noted that males of one of his anophelline populations preferred forming their swarms over the head of a man, and that they would follow him about as he walked slowly around the cage.   This is doubtless totally unconnected with any predatory advantage, since as mentioned earlier, males do not drink vertebrate blood.

Despite all the work done over the course of two years, Bates was unable to discover the ultimate purpose of the male swarming. Was it primarily to stimulate the males themselves, or to attract the females, or was it both? Was the attraction of the swarm an auditory one, a visual one, or both? We still do not know. But the closest Bates came to one of the possible purposes, that of auditory stimulation, was the result of an accident. One of his assistants entered the cage one day humming a rather monotonous Albanian tune under his breath. There was an immediate reaction by the males. Bates then tried it himself. "Swarms in the big cage . . . responded immediately to a low hum, middle C being about the most effective pitch. The swarm would lose its form, many of the males coming and buzzing round the observer's head, the abnormal behavior persisting as long as the hum was maintained. When the humming was continued intermittently for some time, the response became weaker, and after fifteen minutes of trials, there was no response at all, the swarm remaining undisturbed."

It was another ten years before this clue was followed up, and this time the work was done not in Albania, but at Syracuse, New York, at the Cornell University School of Preventive Medicine, by Morton C. Kahn, William Celestin, and William Offenhauser. Again the Rockefeller Foundation aided in the research; the foundation supplied Kahn and his associates with a variety of anopheles mosquito species. They applied the most sophisticated technology to the recording and analysis of mosquito sounds. Most of us regard the whine of a misquito to be a function of its flight, the result of the insect's beating wings moving against the air, like the buzzing of a fly. But Kahn and his co-workers, who may well have believed the same thing before they began their work, were surprised. "In not a few respects," they write, "the sounds of the mosquitoes we have tested are like bird calls. Their variety seems to indicate that they may be in the nature of (a) mating calls, (b) calls of warning or danger, (c) calls of anger and other sounds that are similarly functional." The sounds they recorded were all well within the frequency range of human hearing but far too weak for the unaided human ear to perceive. Among mosquitoes the female, they noted, has the louder and deeper voice. After some time Kahn and his associates were able to distinguish merely by listening to the amplified tapes, without resorting to mechanical analyzers, both the sex and species of the several mosquitoes they studied. The various sounds were produced in various ways— by beating the wings in flight, by rubbing the tarsi against the wings—and they report in addition that there were "certain pure, bird-like sounds, the origin of which we have not been able to determine."

Kahn noted that as it is an odd human who talks audibly to himself, it is the odd mosquito (statistically speaking) who makes sounds without the presence of another insect within earshot. When they did manage to capture the sound of one single female on a tape and then played it back to a group of males, Kahn reports that if the males were of the same species they then "burst into an answering chorus." Moreover, when the call of a female was transmitted to two or three males within the circumscribed space of a small test tube, "it has been observed under the microscope," they write, "that the antennae and hypopygium of the male will turn toward the direction of whence the sound is being transmitted."

One of the female-attractive functions of the swarm may be the amplification of male mating calls. Kahn notes that when a number of insects of either sex were confined and making sounds simultaneously, some of them would vary the pitch of their voice slightly so that the ensuing corporate sound would have a "beat" of resonance.

Though the full range of attractive stimulae produced by the male mosquito swarm is still not fully understood, the sequence of events has been observed and reported when a female is attracted to the swarm. The following description, offered by Bates, refers to an African genus, not one of the European *maculipennis* species, but the behavior is roughly analogous. After a female is attracted to the swarm, one of the males separates himself and "dashes at her." Then "the male, having established contact with the female, the latter almost immediately settles. The male now hangs head downward, suspended solely by his *terminalia* [the external, clasping organs of his genital armament] and may retain this position for over an hour, quivering rapidly and violently throughout."

Among the "longwings" the copulatory contact is apparently established during a "nuptial flight"; the mating couple rises to a considerable height, perhaps 200 feet, whereupon they allow themselves to fall to the ground while continuing to copulate all the while. Swarms of male "longwings" will not form when there are winds of any but a very modest velocity. Their particular requirement of the nuptial flight and the copulatory fall has made it difficult to breed "longwings" in the laboratory.

All the while Bates was involved in these breeding experiments, other ethologists were sloshing about in the marshes of Europe with their pie-pan sluices, seeking knowledge of the mosquito in its other incarnations, as egg, as larva, and as pupa. There were no great differences to be found between members of the various populations in the larval and pupal stages, but the eggs differed considerably.

Mosquito eggs are not the smooth, round, "egglike" entities that one would expect. They are baroque objects, bargelike in shape; oblong, with rounded ends and a round bottom, but with a flat deck topside. Around the "gunwale" there is usually an ornamental frill, and running from the keel to the frill are hollow floats, which may appear either as transparent blisters or in the form of vertical flutings like those found on George III silver teapots. The float serves to keep the egg buoyant where it is maintained atop the water by the surface tension of the frill.

Most of the work done in matching the egg to the adult

mosquito that eventually appeared from within it was done by
two zoologists, Lewis W. Hackett and Alberto Missiroli. Their
work delighted the Linneans and made the segregation of the
various *maculipennis* population into species categories acceptable
to the most conservative taxonomists. By the middle of the
1930's they had "diagnostic criteria" (by which they meant
differences in appearance), so that they could hopefully recognize
differing populations as being different in at least one meta-
morphic stage of their lives. They were now agreed that the
Anopheles maculipennis mosquito populations of Europe should
be classified into seven species. *A. maculipennis*, formerly the
considered chief villain, was rehabilitated. The contribution of
this animal to malaria in Europe was minimal. The *maculipennis*
population was now defined as occupying the Alps in Europe
where it seemed to prefer to feed on the blood of animals other
than man. It went into hibernation over the winter; its eggs
were marked by two black crossbars on a light background and
were found in cool mountain streams. The traditional killer
mosquito of southern Europe was now named *A. labranchae*;
it had very small dark eggs with a small, fluted float, and these
eggs were normally found in sunwarmed, brackish water never
very far from tidal flats. It seemed to feed preferentially on
human blood and died before the winter set in. The other
killer of the eastern Mediterranean was named *A. sacherovi*; its
eggs were a uniform grey color and lacked a float. They were
found in shallow standing water. The animal seemed to feed
almost exclusively on humans and was unable to hibernate; cold
weather killed it.

Toward the end of the decade, another disturbing note was
reported by two American entomologists, Robert Matheson and
Herbert S. Hurlbut. They found that eggs laid by females
belonging to the same population often differed considerably in
appearance, depending on the time of year when they were laid;
winter eggs appeared entirely different from summer eggs.
These were a hibernating species; and moreover, a single female,

at a single laying, might have several quite different looking eggs in the same "raft." This last bit of evidence confirmed the conviction of most zoologists that the traditional Linnean system of classification by appearance was not to be considered a "law of nature," that it was by behavior that the species must ultimately be known, and by sexual behavior in particular.

[5]

Compatibility

The science of biology in the nineteenth century was obsessed with death. If one reads the literature of the period—not Darwin himself, but the writings of his disciples, particularly Herbert Spencer—one is struck by the ludicrous conceit implied in various works; the species were literally created by death. The gaps between the species were supposed to be filled with the corpses of the "unfit." This may be true enough, for all things that live also die; but the interesting question—how diversity came into being in the first place—was largely overlooked, certainly in the popular writings on the subject. Perhaps this insane, back-to-front emphasis in biological thought was due to the culture that produced it. The termination of life—killing— was a perfectly respectable enterprise to engage in and then to discuss at length in print or conversation. The memoirs of the so-called *gentle*men of the period are filled with the most detailed accounts of slaughter; slaughter of animals, slaughter of human beings. But those activities which bring life into being were a totally taboo topic. People obviously engaged in it but they refrained from speaking about it. The one man of the period whose autobiography was much concerned with this side of existence, Frank Harris, became notorious because of what was considered a scandalously misplaced emphasis.

The managers of society were preoccupied with competition. The unwholesome effects of this monomaniacal concern could be seen at every hand, in the frightful conditions of urban slum life, and in the bare-boned poverty of the rural peasantry. In order to quench their consciences, intellectual apologists for the system were eager to seize upon any doctrine that could serve to rationalize their conduct as being in conformity with some "natural law."   To its discredit biology produced that rationalization.   The irreconcilable paradox inherent in this idea that the creative effects of capitalism were somehow produced by the act of destruction is perhaps best encapsulated by the commonplace expression used to denote a rapid speculative triumph—"to make a killing."   The verb and the noun are utterly incompatible with one another in meaning.

This strange bias of mind, the idea that creation depends exclusively on death, that one tree must fall before a new one rises, was not broken until after the *Anopheles* mosquito controversy ended in the middle 1930's. It was obviously not the mosquitoes themselves that produced the change; the controversy merely happened to blow up at the right time.   The values of society were changing and these changed values were reflected in the shift of emphasis in biological speculation.   Killing was no longer seen to be a glamorous and noble occupation.   In the aftermath of the First World War, Europe stank of death.   In addition the Great Depression which followed in the next decade had begun to create doubts even in the narrow minds of the mercantile elite about the virtues of unrestrained competition, doubts about the inevitability of benign effects accruing from this particular "natural law."   Finally, the thought of Sigmund Freud achieved wide currency, and the importance of sexuality as a motivational force in human conduct became generally recognized.   Sex was loosened at last from that matrix of shame and disgust in which it had been embedded for over a hundred years.

With this shift of emphasis, now that biologists were no longer concerned with the benefits obtained by predation, competition, survival under hostile conditions, etc., they became interested in life, in the positive, sustaining side of animal behavior.   Interest now began to focus on animal communication, on animal social organization, and on sexual behavior in general.

The nineteenth century had considered sexual behavior to be activity restricted to males who searched for sexual opportunity as though it were just another of life's commodities, and, like all the others, one in relatively short supply. As a result, they saw animal sexuality as they did almost everything else—as a competitive struggle, a struggle on the part of males to impregnate reluctant females. Hunters blundering into the woods, seeing antlered bucks "locked in mortal combat," and polite ladies who looked over the rims of their teacups to notice the songbirds quarreling on the manicured lawns—all these observers believed that males were competing with one another over the possession of females. In the last twenty or thirty years, ever since patient ethologists had been watching these behaviors with scrupulous attention to the details of sequence, many of these quarrels have been recognized for what they are—territorial disputes. Males quarreled with males over possession of space, not over the possession of females per se, though in many instances, possession of space by the male determined his eligibility from the female point of view. Males could not obtain a female until they had seized a territory.

We human beings, in common with the other more advanced members of the primate order, have lost our dependence on our olfactory sense for information-gathering, and in so doing lost much of our talent for communicating by chemical means. Still, many of our most deeply ingrained habit patterns, such as the social importance among us of food sharing, of dining together as a social ritual, and of taking stimulants or narcotics with our meals—tea, coffee, alcohol—cannot be understood within its evolutionary setting unless this tremendously important universally employed system of communication is viewed in at least some of its multitudes of ramifications.

During the recent integration crisis in the South, chemical communication and its buried relative, the species problem, erupted like some prehistoric social boil over the question of integrated lunch counters. The integration of public transport (when, during rush hour, travelers are pressed into intimate physical contact with one another) took place with comparative ease. But when integration was attempted in eating places, irrational violence exploded on occasion after occasion, for no apparent reason. Here again, the parallels to certain insect behaviors may offer a clue to the underlying motivation. Among insects, ritual food-sharing (as will presently be discussed) serves as the primary communication system by which the society forms itself cohesively as an organism. Among human beings as well, at all times in all societies, the act of eating together forms the basis for creating a societal communion. The act of sharing food with one another seems to be one of the principal bases for creating societies, whether they be of the insect or human variety. It was over this ancient issue of sharing food that human beings in twentieth-century America suddenly found themselves behaving more like insects than like creatures created in God's image and endowed (at least by Linneaus) with sapience.

White Americans did not want to include Negroes within *their* society, and they understood at some infinitely deep intuitive level that if they went so far as to share food with the Negro, they could no longer effectively exclude him from the societal organism.

Among insects, ritual food-sharing is the means by which chemical information is processed and dispersed within the group. A minute amount of chemical information is passed along from individual to individual along with the food particle. These chemicals mechanically alter the behavior of the animal ingesting them. Analogous behaviors are indulged in by humans as they ingest behavior-altering chemicals (tea, coffee, alcohol) along with food in a social setting.

But to begin at the beginning—it would seem as though the reason for employment of chemical molecules for the most primitive (and yet one of the most effective) of all communication systems has to do with the fact that it is not necessary (at least not in primitive organisms) that any specialized receptors be designed to receive the signal. The entire body is aware of the chemical medium, especially in the case of single-celled organisms, in the same way as we are aware with our entire body of heat or cold— or of such a chemical substance as mustard gas, for example.

Since the most primitive existing organisms inhabit the sea, one of the earliest and most curious studies of chemical communication was conducted by two marine biologists, Frank Lillie and Ernest Everett Just, in 1913. One summer night during the dark of the moon they set out from the Woods Hole Biological Laboratory in a rowboat with a carbide lamp swinging from the bow. They wanted to know more about a most peculiar phenomenon, the swarms of minute marine worms which covered the surface of the sea at certain times like a red carpet, rising and falling in the gentle ocean swell. Lillie wrote the report: "The swarming usually begins with the appearance of a few males, readily distinguished by their red anterior segments and their white sexual segments darting rapidly through the water in round paths in and out of the circle of light cast by the lantern. The much larger females then begin to appear, usually in small numbers, swimming laboriously through the water. Both sexes rapidly increase in numbers during the next fifteen minutes and in an hour or an hour and a half all have disappeared into the night." Lillie and Just imagined that this assembly was connected with reproduction. Males and females were obviously swollen with sperm and eggs; but they were unable to observe the mating act under these field conditions, so they captured several specimens and brought them back to the laboratory, putting them in separate containers to see what would happen. They believed it likely that a lunar rhythm controlled both the assembly and the actual spawning activity. They wanted to see whether the females dropped their eggs first, to be fertilized immediately thereafter by the males shedding sperm, or whether perhaps the sequence would be reversed. But they were disappointed.   Nothing happened until, as is so often the case in science, there was a procedural accident. "One day," Lillie writes, "a male was dropped accidentally into a bowl of seawater which had previously contained a female.   He immediately began to shed sperm and swam round and round the bowl very rapidly casting sperm until the entire 200 cubic centimeters was opalescent."   The female had obviously left some chemical trace of her presence in the water, which stimulated the male. But the female had not previously shed any eggs into this water. Lillie then placed a female in the bowl in which the male had shed his sperm, and the female immediately cast her eggs.   Lillie and Just were unable to extract or identify this substance.   They did determine, however, that it was highly species-specific. The worm they worked with was *Nereis limbata*, and when they attempted to repeat the experiment with closely related species of worms, there was no response.   This particular chemical language was understood only by *N. limbata*.   No other worm could comprehend it.   In 1951 a team of German biochemists attempted to analyze this (as they called it) "sex stuff," without too much success.   They did discover that it was a protein substance which oozed out from the entire body of the egg-bearing female.   This stimulates the male into shedding sperm. Chemical elements in the semen stimulate the female into casting her eggs into this floating mass of sperm.   The system is very effective.   It does not require any auditory or visual communication system.

In another related worm, *Platynereis megalops*, this chemical communication system is amplified by a strange behavioral sequence.   In this instance it appears that the males are only attracted to the proximity of the female by this "sex stuff." Ejaculation of male sperm is stimulated by the oral manipulation of the male anal segment by the female. Lillie's associate Edward Everett Just wrote the report on this occasion.   The sequence begins with the female swimming rapidly in tight circles exuding sex stuff.   Males attracted to the general location, directly as they cross the path of a female, alter course, either overtaking or intercepting her. "He then entwines the female," Just writes, "and straightens out, thus clutching her in the twist of his body." He then works his way forward till he has caught her with only one coil of his body behind her head. His tail is free and lashes about. Then "he thrusts his tail into the coil of his body and so into the waiting mouth of the female. The female is quiescent throughout. About six seconds after the female has received the anal segment of the male, the eggs stream from the posterior segments of the female." These eggs are already fertilized, for the female has taken into the mouth the male sperm which has found its way through apertures in the gut to the egg cavity. Just reports that "sections of gravid females just after copulation show sperm among the antennae, in the mouth, in the pharynx, and in the body cavity."

All of this is curious and interesting, but it is relatively unimportant to human beings. It requires an event which affects humans to stimulate full-scale research. Just as it was in the case of the malaria-carrying *maculipennis* mosquito which provided the first unmistakable evidence of "sibling" species and broke open the minds of zoologists to contemplate the entire species problem from a new point of view, so it was another insect plague that opened the door to a new understanding of chemical communication.

During the summer of 1869 a French painter and amateur naturalist named Leopold Trouvelot took a house for the season in Medford, Massachusetts—at 27 Myrtle Street, to be exact. The address has become notorious in the annals of entomology. M. Trouvelot wanted to do some illustrations of silkworms in various stages of their development, and to that end had imported several eggs from Europe. He left them on a table in his workroom and waited for the larvae to emerge. It must have been a warm summer, and the table on which the eggs were located must have been near a window. Perhaps the rustling of a curtain blowing in the wind brushed them from M. Trouvelot's worktable and out the window into the yard.   M. Trouvelot missed the eggs on the afternoon of their disappearance and made efforts to find them, for among the eggs were several belonging to a European moth, *Porthera dispar*, known in England as the gypsy moth, perhaps because the color of the wings of the male imago resemble the color of tanned, windburned skin.   Knowing full well this moth's European reputation as a destroyer of leafy trees, M. Trouvelot "gave public notice of [the eggs'] disappearance."   It was a pity that the public took no notice at the time.

Within two years of this event, a vigorous colony of the moths had established itself on Myrtle Street and the citizens of the town of Medford found themselves living in a nightmare.

There was a public hearing sometime later at which citizens spoke of their impressions of that summer of 1871.   One Mrs. Belcher reported as follows: "My sister cried out one day, 'They [the caterpillars] are marching up the street!'   I went to the front door and sure enough the street was black with them coming across from my neighbor, Mrs. Clifford, and heading straight for our yard."

A Mrs. Spinner stated: "I lived in Cross Street . . . and in June of that year [1871] I was out of town for three days.   When I went away the trees in our yard were in splendid condition and there was not a sign of insect devastation upon them.   When I returned there was scarcely a leaf upon the trees.   The gypsy moth caterpillars were over everything."   A neighbor of Mrs. Spinner, Mr. D. M. Richardson, said: "The gypsy moth appeared in Spinner's place in Cross Street and after stripping the trees there, started across the street.   It was about five o'clock in the evening that they started across in a great flock and they left a plain path across the road.   They struck into the first apple tree in our yard and the next morning I took four quarts of caterpillars off one limb."   Mrs. Hamlin continues: "When they got their growth, these caterpillars were bigger than your little finger and would crawl very fast.   It seemed as if they could go from here to Park Street in less than half an hour."

Mr. Sylvester Lacy reported that: "I lived in Spring Street . . . and the place simply teemed with them and I used to fairly dread going down the street to the station. It was like running a gauntlet. I used to turn up my coat collar and run down the middle of the street. One morning in particular, I remember that I was completely covered with caterpillars inside my coat as well as out. The street trees were completely stripped down to the bark. . . . The worst place on Spring Street was at the houses of Messrs Plunket and Harman. The fronts of these houses were black with caterpillars and the sidewalks were a sickening sight, covered as they were with the crushed bodies of the pest."

The destruction in Medford was horrible. The landscape looked like a burned-out battlefield. In their desperation the insects had devoured even the grasses down to the bare dirt. The Massachusetts National Guard was called out to drench the countryside with Paris Green, an arsenic compound. Artillery caissons were converted into spray wagons, but the moths continued to flourish.

The plague spread very slowly. Female moths, though they did not appear markedly different from the males, are flightless. When they emerge from the pupa case, they can crawl only a few feet away from the spot where they are sought out by flying males and fertilized, so the new generation of eggs is laid not far from the previous spot.

Dr. Charles Fernald, the resident entomologist at the Massachusetts State Agricultural Station at Hatch, was the first professional zoologist to become involved with the moth. He was not at home when the first delegation of citizens from Medford appeared on his doorstep with specimens of the caterpillar and the moth. His wife searched vainly through his collection of American moths in an attempt to identify the animal, and it was finally his young son who discovered Linneaus' original description of the moth in a European book. Fernald found the havoc caused by the moth utterly incredible until he performed some tests of his own with lettuce leaves and found that a fully mature caterpillar would consume well over ten square inches of foliage in one twenty-four hour feeding period.

He wondered how the male moths located their flightless females. Was it random accident? The vernacular name for the moth in France was "Zigzag," which described accurately their erratic, wavering flight. Fernald suspected there might be some truth to the folklore account that it was a scent exuded by the female which attracted the males, and that their lumbering flight pattern had evolved in response to this need. In traveling from one point to another the moth lumbers wildly off course, to and fro and up and down: it would thus more likely encounter air-borne scent traces than if it flew sharp and straight like a swallow.[1]

Fernald devised a trap, a wire-mesh box with a funnel entrance, which would admit insects but not permit them exit, and using live females for bait, he attracted numbers of males into it. He began using these traps as a gauge of infestation in any given area. By placing the traps in an open location and leaving them for several days, he could get an impression of the population density from the number of males captured. By this time the plague of caterpillars had spread from Medford to other Massachusetts towns, probably carried by travelers such as Mr. Lacy who dreaded going to the railroad station covered with caterpillars, but went nonetheless.

In all of this work by Fernald and in the work to follow, there was very little of the philosophical speculation which so marked the studies stimulated by von Uexküll or even Marston Bates working with his anopheles. It was all eminently practical. By 1915 the moths had spread as far north as Maine, and in

[1] Since this original work with the gypsy moth, chemical sex attractants have been found to operate among so many moths and butterflies that it is now suspected to be universal. The wavering flight typical of these animals does not aid butterflies whose sex stuff operates at short range (they locate one another visually from afar), and in the case of butterflies its adaptive purpose would seem to be an evasion tactic, to escape predation by birds. It may also operate in this way for the gypsy moth.

August of that year entomologists had become curious about the range of this sex attractant. Traps containing live females were placed on several uninhabited islands off the coast of Maine, in Casco Bay. In one trap placed on Outer Green Island, better than two miles from the nearest moth, and left there for three weeks, two males were found when it was retrieved. In another trap placed on Billingsgate Island off the coast of Cape Cod, four males were found, though this trap was two and a quarter miles from the nearest point of infestation.

To assure themselves that these males had actually flown the distance and not been blown by winds into the general vicinity of the traps, follow-up tests were conducted in a closed, airstill room twelve by fourteen feet, and the moth's flight speed was timed at approximately 150 feet per minute. Virgin males rarely flew uninterruptedly for much more than a mile before alighting, but a small number of sexually experienced males flew constantly for several days, covering well over two miles before eventually collapsing in death.

This is curious in that prior sexual experience would seem to be a powerful motivating force for males. Fernald timed a number of gypsy moth copulations and reported that they ranged in time from twenty-five minutes to three hours and eighteen minutes, with the average time being about one and a quarter hours. He writes that "after mating, the male is quite stupid but in about one-half hour regains his normal activity" and is quite eager to mate again.

The next episode in the annals of the study of the sex stuff of the gypsy moth is almost comic in its foolishness. For the amount of sophisticated effort that was lavished upon obtaining the information, very little was finally learned. But the gigantic power of the United States Government was now bent upon the project, and the elephant labored to produce the mouse. By 1927 the gypsy moth was no longer a regional problem; it had become a national concern, and the United States Department of Agriculture was asked to aid in its control.

It was imperative now for the Government to obtain a census of the moth. Fernald's system of using live females as bait in traps was considered too dangerous: if, through some inadvertence a fertile female should escape, the horrors of Medford might be re-enacted somewhere else. Two government entomologists, Charles W. Collins and Samuel F. Potts, were assigned the task of devising an extract of female scent to be used for bait. After dissecting thousands of moths and baiting traps with sections of tissue taken from various parts of the female anatomy, they finally discovered (not to anyone's great surprise) that the scent was produced in the tissues immediately surrounding the female genitalia, particularly the area immediately surrounding the opening of the copulatory pouch. But though the Government contracted the services of several eminent university chemistry professors, no one was able to isolate and identify the chemical responsible. However, the Government is obsessed with the establishment of standards and specifications, and the notion of utilizing female genitalia (even if only that of a moth) as official United States Government equipment must have been repellent to the authorities in charge of the program, for thirty years were spent in the attempt to isolate the chemical in its pure form. Finally in 1960 three chemists, Martin Jacobson, Morton Beroza, and William A. Jones, working at the Beltsville, Maryland, United States Government Agricultural Research Station, succeeded. They had dissected the genitalia of well over half a million female moths in order to determine the official specifications for the scent bait to be used in United States Government gypsy moth census traps. For the record, the name of the chemical is dextrorotatory 10 acetoxy-1-hydroxy-cis-7-hexadecene.

The search-and-copulate behavior of the male gypsy moth, which is stimulated by this chemical substance, is called in the language of ethology a *releaser mechanism*. While many biologists consider it a violation of reality to construct mechanical models for the behavior of living things, mechanistic language does often offer a convenient way of correlating disparate events, which appear to have a causal relation to one another. With this disclaimer one can compare a releaser mechanism to the trigger of a gun. When a marksman presses the trigger of a gun, the only control he can exercise over the subsequent behavior of the bullet is directional. He can point the barrel here or there. The actual behavior of the bullet itself, its velocity, is determined by the cartridge that encloses it—the amount of powder and its speed of combustion.

But a releaser mechanism *cannot* be compared to the action of an archer shooting an arrow. The archer chooses his arrow, be it sharp or blunt, long or short. This is a volitional, judgmental choice based on expediency. The archer only fits the arrow to the bowstring immediately before shooting it, while the bullet has resided in the gun for some time before the trigger is pulled. For the gunman there is no choice of bullet type. The archer is also able to maintain complete and continual volitional control over his activities until the arrow has left the bowstring, while as soon as the gunman pulls the trigger a preplanned series of events occurs over which he has absolutely no control whatever; the hammer strikes the primer which ignites the powder, etc.

The action of the gypsy moth is preplanned like the action of a gun. "It is certain," wrote Charles Darwin, "that there may be extraordinary mental activity with an extremely small absolute mass of nervous matter. Thus the wonderfully diversified instincts, mental powers and affections of ants are notorious, yet their cerebral ganglia are not so large as the quarter of a pin's head. Under this point of view, the brain of an ant is one of the most marvelous atoms of matter in the world, perhaps more so than the brain of a man." The means by which such incredible miniaturization was accomplished is more apparent to us than it was to Darwin. The brain of any insect, whether it be solitary like the gypsy moth, or social like the ant, is (in the current jargon) "preprogrammed" with a certain number of stereotyped behaviors. The analogy between insect brains and modern computers is not at all farfetched; the organism is quite compatible with the machine in this case, and ants are far more readily adaptable to life in the computer age than humans, for they both receive information in the form of a binary code. This is the basis of chemical communication; the signal is binary—that is, two-pronged. It is either on or off. To resort to the analogy of the gun, the signal either actuates the trigger releasing the preplanned sequence of activities, or it does not actuate the trigger. The most ambitious extent of a binary code comes when it stimulates an alternation of events, that is when it actuates the trigger or *inhibits* the actuation of the trigger (by setting a hypothetical safety catch).

After the breakdown of Linnaen classification which followed on the malaria mosquito controversy, promoters of the New Systematics became fascinated with the problem of sexual compatibility—within a large general population, which male will mate with which female. As it is with human sexual compatibility, so it is with animals—basically a matter of communication.

In ordinary human affairs, a breakdown in communication is generally considered catastrophic. But from the point of view of the taxonomist something new and different can only begin to occur just at that moment when communications finally do break down for good, when a splinter portion of the population finds itself so alientated from the parent group that it turns away and in upon itself, and in the process develops some new and special characters.

Had it not had such a long and pernicious history of theological misuse, the word excommunication might well have been the term taxonomists chose, instead of the one they finally used to describe this process of alienation—the isolating mechanism.

Whenever there arises a special language, there are some who are isolated from the communal life of those who speak it, whether that language be of the late Beethoven quartets, or of *Finnegans Wake*, or of the special theory of relativity. This exclusivity on the part of that splinter which pursues its specialized concerns does not make any of us the poorer. On the contrary, we are enriched by this diversity of the human mind and spirit.

In the 1940's when the promoters of the New Systematics sought out a vehicle to test their hypotheses, their choice centered upon that animal whose sexual behavior had been the most rigorously documented. It was an insect; it had served as the prime organism for inheritance studies ever since 1909, when a zoology professor named Thomas Hunt Morgan, then working at Columbia University, became interested in Gregor Mendel's theories of heredity. Rather than choose, as did Mendel, a botanical organism for study, Morgan selected an animal organism convenient to him, the so-called vinegar fly (or fruit fly), *Drosophila melanogaster*. It was small, hardy, easily adapted to confinement, and had a relatively short reproductive cycle, so a considerable number of breeding generations could be observed within a comparatively short span of time. Morgan's choice proved to be a most fortunate one. Shortly after he began working with these animals, it was discovered that in the entire class of two-winged flies (*Diptera*) there exists, during their larval stage, giant chromosomes in the cells of their salivary glands. These chromosomes inhabit huge nuclei; they are several hundred times larger than the chromosomes contained in the reproductive cells, the sperm and ovum. And yet they are identical to them, since every cell of the body contains chromosomes identical to those of every other cell. Under the microscope, when properly stained, these chromosomes look like angle worms, being composed of bands or striations. It was hypothesized that certain genes typically occupied special locations on these bands, and the mapping of gene locations—that band on the enlarged chromosome which contained the gene which controlled such characters as eye color, or wing shape, or hairiness—became one of the professional games of geneticists (by which they obtained status, recognition, and promotion) for two generations. Since 1909, therefore, fruit flies have been the object of the most fanatic concentration by zoologists, and it is no wonder that when experiments designed to test the credibility of the New Systematics were devised, fruit flies were the animal to which everyone turned as the experimental vehicle.

Though it is small sized the fruit fly is by no means a simple organism. It is highly evolved both in structure and behavior. Its courtship and copulatory rituals approach in complexity those of many mammals. Herman T. Spieth of the College of the City of New York has written an excellent account of the sex act as performed by the fruit fly. He studied his flies by enclosing them in vials or test tubes and then observing them under a low-powered microscope. As soon as he inclosed several flies of both sexes in one of his tubes, they began moving about, tapping one another with their forelegs. If one of the males encounters a female, Spieth writes, "the male moves to her rear, extends his proboscis and proceeds to lick her genitalia. At the same time he extends his forelegs. . . . Licking and rubbing typically occur together and extend over relatively long periods of time with only occasional short pauses.

"A receptive female accepts [this courtship] by spreading the wings outward and upward . . . at the same time spreading the genital plates. The male then mounts and typically inserts just as he mounts. . . . During copulation, he intermittently rubs the sides of the female's abdomen with his [middle] legs in rapid bursts of movement.

"The female is typically quiet during copulation, but near the end, she becomes restless, kicks vigorously with her hind legs and walks about. The male determines the time of termination, and typically he withdraws and then dismounts, when he is ready immediately to court again and often courts the female with which he has just copulated."

Spieth timed copulations with a stopwatch and reckoned that the time required for copulation varied from one minute and fifty-one seconds to seven minutes and thirty seconds, with the average time for the several species he studied running somewhere about three minutes. Not all males and females were equally interested in courtship. "Often," Spieth writes, "it was observed during the course of these experiments that a particular individual male was extremely aggressive in his courting behavior while another male in the same vial merely sat, preened, and fed. Likewise some females accepted readily while others in the same experiment steadily refused even when courted vigorously and persistently."

Since it was the male Drosophila which initiated these sexual encounters, Spieth was eager to discover which of the several stimuli involved, was actually the critical one—which produced the negative signal which might terminate the courtship prior to the act of mounting and insertion. Some of his males interbred freely with females from other species. This did not necessarily invalidate the New Systematics since the behavior *in the wild* was the defining factor. And in the wild food preferences (and many insects meet one another socially while feeding), activity rhythms, and geographical location might make it statistically highly unlikely that a male and female of differing species ever meet in sufficiently close proximitiy as to permit courting.

When he discovered patterns of aversion, Spieth was curious whether he could determine what signal turned the behavior on or off. He found that the males of the *Drosophila virilis* species were highly selective. "In crosses involving *D. virilis* males and females [of other species] the males almost invariably broke off the courtship at the tapping stage. At the same time, however," he writes, "the males courted each other vigorously. The *D. virilis* males never showed signs of sexual excitement when in the presence of females of these [other] various strains, but if at the end of the observation period *D. virilis* females were introduced into the observation cell, then the males would become highly excited sexually. Even these sexually excited males with greatly lowered thresholds would not court foreign females."

Though he could not be sure, Spieth believed it was a sense of taste, a chemical signal that produced the sexual appetite. "On the part of the male," he writes, "the stimulus to mount seems to be received directly by his proboscis, and to originate in the genitalia of the female. . . . Numerous times males were observed standing behind solitary uncourted females that gave the accept- ance response. . . . In no case did any male evince any response to such a receptive female. . . . Often when courting, the male would be rubbing and vibrating, but instead of licking the female genitalia, would have the tip of his proboscis against the abdominal sclerites [plates]. If with this set of conditions pertaining, the female gave an acceptance response (as she often did) the male never mounted unless he subsequently reached the genitalia with his proboscis."

"In a normal courtship it was clearly observable that after the female gave the acceptance response, the male's proboscis was then pushed into the opened genitalia which he continued to lick for a short but definite period before he actually mounted."

While Spieth was working at City College observing the sexual stimulation of male fruit flies, Ernst Mayr, one of the original formulators of the New Systematics, was working at the Ameri- can Museum of Natural History, also in New York, watching females. Mayr devised a somewhat more exclusive experiment. He suspected that the female acceptance response of spreading the wings and genital plates must be triggered by scent, since often the male stands off at some distance when the acceptance signal is given. He therefore decided to concentrate his experiment "on the olfactory sense which in *Drosophila* is located in scent organs in the third segment of the antennae. Removal of the antennae deprives *Drosophila* of its sense of smell."

Mayr feared that the operation of removing antennae from female flies could sufficiently debilitate them that they might be disinclined to accept the sexual advances of males on account of lowered general health and well-being. Therefore he performed an even more drastic operation on a second set of females which he used for a control group. He operated on this second set of females by removing their proboscis, an operation which left them unable to eat or drink and from which operation they would perish within twenty-four to thirty hours.

He then enclosed some normal females and some antennae-less females in a test tube with males. "The males display actively to both kinds of females," he writes, "perhaps even more to the operated flies, which appear slightly more lethargic, than to the other. In spite of frequent genital contacts, no copulation with antennae-less females is completed. The normal females copulate after four minutes, twenty-three minutes, and twenty-eight minutes." When he enclosed normal females with proboscisless females in a tube with males, Mayr was able to note no significant differences in acceptance of copulation.

Mayr recounts a conversation he had with Spieth while both men were engaged in this work, during which Spieth recounted how the male approaches the female afoot, but buzzing his wings vigorously. "Spieth believes," Mayr writes, "that the fluttering wings serve like a reverse propeller throwing an air stream toward the courted female. This air, containing the scent of the male stimulates the female and increases her receptivity. If the male belongs to a different species . . . there will be less stimulation. A reduction in the amount of olfaction is also produced if the wings of the male are removed."

It is obvious from Spieth's account that other stimuli, particularly tactile, are actively involved in the communication sequence. The more we learn about these rituals of behavior, the more incredible the parallelism of behavioral evolution becomes. Though we are very little aware of it on the conscious level, tactile communication still remains tremendously important to us humans, not only in the communion of the sex act but also in general terms.

This is particularly well recognized by politicians who will undergo discomfort and sometimes risk personal injury in order to "shake hands," establishing tactile communication with their constituents. Even in this day of verbal and visual communication saturation the intimate force of tactile communication still persists. Politicians will undertake hand-shaking tours because it has been established, at least to their pragmatic satisfaction, that the effects of this intimate, person-to-person tactile communication are more permanently affecting than verbal communication at a distance, through print or television. Politicians believe that the act of touching hands effects a permanent commitment, and that a hand shaken is a vote in the ballot box.

In human societies all over the world, some sort of tactile communication invariably accompanies verbal communication in the greeting ritual. The implications seems to be that one can detect possible deception in the contact of flesh with flesh, when one might not be able to hear it in the spoken words.

In the work with Drosophila, Spieth and Mayr were concerned with the evolutionary role of excommunication as a stimulator of diversity. A more conventional view of communication as the means by which separate individuals may be organized into a cohesive social framework was being studied at the same time by two German entomologists, Peter Karlson and Adolph Butenandt, who were working at the Max Planck Institute in Munich. They recalled a hypothesis originally proposed by William Morton Wheeler, the great Harvard entomologist, in 1910. Wheeler imagined that insect societies had more in common with multicellular organisms (which he called *persons*) than they did with the politicians' view of human society as in interplay of competing power blocs. He did not like the usage of the term organism because, as he wrote, "the organism is neither a thing nor a concept, but a continual flux or process, and hence forever changing and never completed." He wrote, however, that he found the notion of the organism useful in his study of ant societies "and," he continued, "as I have repeatedly found its treatment as an organism to yield fruitful results in my studies, I have acquired the conviction that our biological theories must remain inadequate so long as we confine ourselves to the study of cells and persons and leave the psychologists and sociologists to deal with the more complex organisms.

"Indeed, our failure to cooperate with these investigators in the study of animal and plant societies has blinded us to many aspects of the cellular and personal activities with which we are constantly dealing. This failure, moreover, is largely responsible for our fear of the . . . metaphysical, a fear which becomes the more ludicrous from the fact that even our so-called 'exact' sciences smell to heaven with the rankest kind of materialistic metaphysics."

As might be expected, Wheeler was at the time widely attacked for having an insufficient fear of metaphysical speculation, and therefore indulging himself in this dreadful sin. By 1918 Wheeler had arrived at the point where he believed that insect societies were kept orderly and cohesive through the practice of what he called "trophallaxis"—ritual food sharing, and that this was analogous to the way various cells were nourished in a multicellular "person."

It would seem, upon reading the terse unspeculative report of Jacobson and his associates, that they had labored to synthesize the gypsy moth sex attractant in order to produce a useful catalogue item for the United States Department of Agriculture and nothing more. Karlson and Butenandt, though they were bent on a similar task—the isolation and synthesis of the sexually-attracting secretion of the silkworm moth—were out for bigger game. "The attractant of a moth," they write, "is produced and secreted by certain glands just as is a hormone; even the minutest amounts cause a reaction in the receptor organ (antennae) of a male which induces the male to copulate. But contrary to hormones, this substance is released to the outside and not into the blood. It does not serve the humoral regulation inside the organism, but rather acts upon individuals." Accordingly, they proposed to name this class of substances with a special term, *pheromone* (the suffix derived from the Greek *pherein*, meaning to carry), in order to distinguish it from ordinary chemical molecules, which like the odor of smoke or camphor, might provide the receptor animal with information and stimulate activity, but which are not primarily designed as a communication *system* between conspecific individuals.

In the gypsy moth the scent is produced by scattered cells located in the genital tissues of the female.   As with the smell of human sweat, the gypsy moth can exercise no control over this scent.   In the silkworm moth this sex attractant has undergone an evolutionary elaboration.   The cells are localized within a gland and this gland is capable of being exposed or retracted, thus giving the silkworm moth a measure of control over the display of her sex scent.

It has been known since 1902 that the honeybee also possesses a scent gland in the abdomen.   The honeybee's gland, named after a Russian entomologist, Nicholai Nasanov, who discovered its presence but not its function, represents an even more highly advanced elaboration.   The cells have migrated up the back of the abdomen from the genital area, and are invisibly inclosed in the overlap between the last two plates of abdominal body armor. When the bee downcurves its abdomen into an arc, thus spreading the plates, this gland can be seen appearing on the insect's back as a narrow whitish membrane.   Superficially it would seem to be part of the plate itself.   A British apiarist named Frank W. Sladen is credited with discovering the function of Nasanov's gland in that year, 1902.

He noticed that as foraging bees return to the hive after an expedition, they pause for a moment on the threshold, curve their abdomens downward, thus exposing Nasanov's gland, and buzz their wings.   As a considerable number of bees were performing this ritual, Sladen noticed a peculiar and distinctive scent.   Several days later, while dissecting several bees, upon exposing this gland, he again observed the scent and remembered it.

Of all the social insects, the habits of bees are best known to man.   In some parts of Africa, to this day bees are the only animals systematically domesticated by man; and it is quite likely that they were one of the first, if not the very first wild animal that many primitive peoples first "tamed" and made use of. For as long as man has known about them, they have been a symbol for him of orderly, useful social management. On the lid of the sarcophagus of one of the kings of Lower Egypt—who was interred in the year 3633 B.C.—is carved the symbol of the honeybee, apparently the king's idea of an appropriate personal emblem for a wise ruler.

It is believed that bees as animal forms evolved from wasplike insects some millions of years ago. Of the 1,500-odd species of "true" existing wasps, about 95 per cent live solitary lives. The remaining 5 per cent include such social wasps as paper wasps and the familiar yellow-jacket hornets. The reverse is true of bees. While there are a great number of solitary species, the great percentage live social lives as members of a bee colony.

Nothing seems ever to have escaped the attentive eye of Karl von Frisch; to him goes the credit for establishing at least one function for the Nasanov gland—that of producing a social attractant scent. There is no need among worker honeybees for a sexual attractant: in the first place they are all females, and in the second, their ovaries are normally atrophied so they perform no sexual roles whatever. In 1919, shortly after he began working with honeybees, von Frisch noted that once his training boxes containing bowls of sugar water had been visited by bees from a hive, other bees were attracted to the same box. It was this observation that put him on the track of the communication dance. However he was puzzled by the fact that the bees seemed to prefer the identical box visited by their fellow workers, regardless of whether or not he put additional boxes right alongside it. At this early stage of his work von Frisch was interested in the color discrimination sense of bees, and this social preference factor disturbed his statistical records. He immediately suspected a scent trace of some kind, and proceeded to devise a simple procedure to test it. He trained bees to accept sugar from a box; then, as soon as the box had been visited a number of times by foragers from a given hive, he added identical boxes alongside. The original box received an overwhelming number of additional visits from bees of the same hive. Von Frisch then repeated the experiment, but this time he trapped the foragers and coated their abdomens with shellac, making it impossible for them to extrude their Nasanov glands. He then found that when he added more boxes, all the boxes in the same general location were visited on a random basis by foragers from the hive; there was no scent marking remaining on the original box which disposed bees to prefer it over the others.

Von Frisch also discovered, however, that the duration of this scent was short-lived. After a ten-minute interval, scent-marked training boxes appeared to lose their scent attractant. The bees did not, apparently, deposit any of their Nasanov substance directly on to the material of the box; they merely exposed the gland in the air, letting the scent molecules disperse. After ten minutes, this dispersal was sufficient so the scent no longer served as an attractant.

The sequence in which we perceive data has much to do with our understanding of it. This is the chief charm of that narrative literary form, the detective story, in which the scraps of reality which lead to the solution of a problem are all there (at least in an honest detective story) but are placed as far out of sequential juxtaposition as the ingenuity of the author will allow. This absence of sequential juxtaposition defers the mind from seizing the pattern presented by the evidence. The same lack of sequence in the various discoveries regarding chemical communication deferred for many years our understanding of its importance.

At first the sex attractant of the gypsy moth was understood as a practical tool for controlling an insect plague. Then it gradually became apparent that it played a role, not only in this particular moth but among animals generally, in the speciation process, in the evolution of diversity. It was, in effect, a language of love that could only be comprehended by compatible lovers. It served as a means by which individuals could preserve and maintain, intact, the corporate body of their collective experience. It served as a means of isolating this population from the contamination of other, similar animals—animals that had suffered differing experiences through living different lives, engaging in different activities, and coming to know the world differently through differing *umwelts*. The communication system thus marked out the boundaries of the path of destiny.

Now, as a result of the gypsy moth discoveries, and Sladen's and von Frisch's discoveries of the social attractor of the analogous gland in the honeybee, one could discern nature's pattern of evolutionary refinement. In the honeybee the chemical signal worked to fragment the population into even smaller populations than that of the species. For social insects it began to appear as if this chemical communication isolated the colony from all other similar colonies.

Generations ago, apiarists knew that bees will repel intruders to their hive, kill them, and—even though a colony might be reduced to a state of hysterical disorganization by the loss of a queen—if an intruder queen is placed in the hive, the residents will very likely sting her to death. As far back as 1815 German apiarists had contrived a system for introducing foreign queens into a queenless colony which seems somehow terribly consonant with the German national character. They knew that bees are attracted by pure sugar, but that a protracted diet of pure sugar leaves them in a staggering, stultified state, very similar to drunken humans. Before putting a foreign bee into a hive, the Germans installed her in a small wooden block, which had been pierced through by holes from various sides. These holes were blocked up with rock candy. Then this box containing the queen was placed in the hive. Since the queen's mandibles are inoperative —she can only take specially prepared food—the queen could not eat the candy from the inside. But workers from the outside could and did eat their way through the candy and into her presence. But the sticks of rock candy which plugged the holes were several inches long and it took the worker bees several days to perform the task of penetrating through to the queen, by which time they were in such a state of intoxication that they received her with every sign of welcome and good fellowship.

The presumption, of course, was that any bee, whether queen or worker, lacking the scent of the hive would be repelled, and that the colony scent attractant operated in much the same way as species scent attractants, but it took a British apiarist, Ronald Ribbands, to prove this conclusively. Ribbands trained marked bees from two separate but otherwise identical colonies to accept syrup from bowls within an open-top box. As von Frisch had discovered, bees merely mark the air with scent—they do not mark the actual sources of food—and so Ribbands, hoping that he could contain the scent within a stipulated area, not allowing it to be wafted around by the wind, placed his sugar bowls within protective walls. He kept moving these dishes and their enclosing boxes closer and closer together until at last they were adjacent—so much so as their walls would permit—and there was no mingling of bees from the two colonies. When the experiment was repeated with a shellac coating over the abdomen which prevented the scent from being dispersed from Nasanov's gland, mingling ensued.

Though they densely populate any given area at certain times of their life cycle, neither Lillie's marine worms nor Fernald's gypsy moths could be considered social animals—they are essentially solitary in that there is no communal purposiveness in their behavior; they do not build a communal nest, they do not co-operate in the foraging or distribution of food, or in the rearing of young, or in any other purposeful way. Chemical communication in the form of sex scent serves, in those examples, only to bring potential sex partners from within the same genetic population into contact with one another and to isolate this population from penetration by members of other genetic populations.

Ribbands had now proved that an evolutionary elaboration of the sex scent, as it was produced by Nasanov's gland in the honey-bee, served a narrower isolating and attracting function. It served to maintain and isolate the communal identity of members of the same social population. And, as this foraging experiment proved, it provided a mechanism by which co-operating members of a community, engaged in a task for the welfare of the community, could be attracted to one another and to the performance of the task—namely, food foraging.

Immediately, larger questions about the larger potential role of chemical communication in the organization of insect societies came into Ribbands' mind, but his first task was the determination of how this scent, unique to the hive, was produced by the hive. Ribbands' first clue came from a paper written in 1924 by a German entomologist named Wilhelm Jacobs, who described having dissected Nasanov's gland and finding it to be composed of a dense concentration of specialized cells. Jacobs also described other cells, seemingly identical to those in Nasanov's gland, scattered widely over the honeybee's body. He believed that these cells might exude waste products of the bee's metabolism, operating analogously to human sweat glands.

To test Jacobs' hypothesis, Ribbands, in 1952, separated the bees of a large colony into three sections. One of the sections he dosed with black treacle, the others he left alone. In a mingling test he discovered that bees belonging to both the untreated sections of the hive mingled freely together, while those from the section treated with treacle behaved as if they belonged to a different colony entirely.

In most natural locations where bees exist, there are a multitude of pollens and nectars to choose from, and it would seem most unlikely that each forager bee would ingest exactly the same mix of pollens as would cause her to smell exactly like her neighbor unless Wheeler's old contention of *trophallaxis* (ritual food sharing) provided all the members of the colony with an identical diet.

To test this hypothesis, Ribbands trained six marked foragers from a colony containing 24,600 bees to accept twenty milliliters (about a teaspoonful) of syrup, which had been dosed with radioactive phosphorus. It required three hours and over 370 separate visits before these six bees had transported all the tracer-marked syrup. Ribbands took samples of the hive population from various worker castes every five hours, and after twenty-nine hours measured the entire membership of the colony. By this last measurement *better than 76 per cent of the members were radioactive*. Some castes contained less radioactivity than others, and drones (who perform fewer social functions than any other residents of the hive) had the lowest percentage—fewer than 2 per cent showed any sign of having partaken of the syrup. But of the eighty-five larvae which were exhumed and measured, 100 per cent carried the radioactive trace.

This discovery by Ribbands, that chemical information possessed by a relatively few members of a colony at large is distributed until it becomes the possession of the majority, set another British apiarist Colin Butler to wondering what additional chemical information, if any, might be passed from bee to bee by ritual food sharing. For as long as man had been observing them, man had been fascinated by the orderly functioning of insect societies. Most social animals communicate by means that are familiar to humans, through the mediums of sight and sound, through vocalization and gesture. Von Frisch had demonstrated that gesture (the ritual dance) does carry information about food sources to honeybees, but what about ants and termites? Anyone who has ever observed them at work is struck by the fact that the furious pace of activity seems undirected by any visible communication passing from one animal to another. An entire flock of crows will be stimulated to flight when alarm cries are sounded by one of its members; a herd of antelope will respond alertly to the ear twitching and wind sniffing by one of its members. The human observer can empathize with such behavioral communication which dispose an entire congregation of animals toward a collective activity. But insects conduct themselves with what seems to humans to be an utterly paradoxical solitary independence, which leads most mysteriously to the fulfillment of some corporate objective.

Butler decided to test for honeybee behavior that might be stimulated by chemical information passed via food sharing from one bee to the next. Describing the behavior he intended to study, he wrote: "When a colony of honeybees loses its queen, the worker bees soon become aware of this fact (often within thirty minutes or less) and the behavior of the colony as a whole tends to change from a state of organized activity to one of disorganized restlessness. If the hive of a colony that has recently lost its queen is opened, many of the bees expose their scent-producing (Nasanov) glands and fan currents of air over them with their wings—thus producing the 'roaring' sound which is associated with queenlessness. Within a few hours, a more definite sign of queenlessness becomes apparent as one or more worker brood cells containing young female larvae will usually have been modified by the bees to form emergency queen cells. The larva in one of these cells is destined, all being well, to become the new queen of the colony. If the state of queenlessness persists, often the ovaries of workers which are normally atrophied are now stimulated into growth and production."

The comparatively slow pace of the communal reaction and the fact that the typical "roaring" had perhaps something to do with the distribution of emanations from the Nasanov glands suggested to Butler that a pheromone, an external hormone, might be involved in this communal response. Since what was known about chemical communication systems suggested that they transmitted information via a binary code, Butler hypothesized the following sequence: If the absence of the queen-chemical produced certain behaviors, then her presence, and the presence of this chemical inhibited these behaviors. The logic may seem odd when related to human behavior, but one could say about humans that if the absence of oxygen produces suffocation, then the presence of oxygen inhibits suffocation.

Suspecting, from the delay in the communal reaction to the presence or absence of the queen, that chemical information about her presence or absence was transmitted along with food particles from bee to bee in the ritual act of food sharing, Butler's first step was to exclude other factors which could conceivably be operative.   In one adroit experiment he proved that neither the sight of the queen in the flesh, nor sounds made by her, nor a generalized scent exuded by her which might have pervaded the entire hive, informed the colony of her presence.   He enclosed the queen in her own colony within a double-walled cage of wire mesh, through which no other bee could make physical contact with her.   Within a matter of hours, though the members of the colony could see her, hear her, and smell her, the colony reacted with the classic symptoms of queen loss.

Butler's next step was to begin a detailed study of the various kinds of physical contact exchanged between the queen and the members of her colony.   The queen is normally surrounded by a suite or an entourage.   These words are the technical terms used to describe this social arrangement—a refreshing change from the usual run of bastardized Latin generally employed in scientific usage.   This entourage generally contains ten or twelve nurses.   Butler wished first to determine whether these nurses composed a special caste within the colony, or whether any worker bee could perform nurse duties.   To this end he removed the queen with a forceps, placing her in various random locations within the hive.   Each time this was done, the suite disbanded as the queen was removed, the nurses going about other worker tasks immediately, while at the new location a new suite formed itself around the queen immediately.   In order to gather harder statistics (the kind that impress other scientists), he repeated the basic experiment, except that instead of using crude forceps he devised a complicated, electrically-powered transport cage, which moved the queen about the hive according to a predetermined pace.   Using this procedure, he demonstrated that information concerning the presence of the queen was transmitted via the nurses who comprised her entourage to other bees via ritual food sharing, and from these others to still others, until,

like a chain letter, the number of bees receiving a chemical trace
of the queen's presence, multiplied geometrically.

The suite of the queen bee surrounds her completely. The
bees directly in front of the queen feed her continuously. Those
bees to her rear and sides stroke and caress her with their antennae
or lick at her with their proboscises. Butler discovered that
those bees who caressed the queen with their antennae then
touched antennae with other members of the colony when they
exchanged food. Those who fed the queen took in this queen
substance via their proboscises and distributed it in the same way
to others during food sharing. In 1962, ten years after Ribbands
had stimulated his curiosity, Butler finally isolated the queen bee
substance. He found that it was produced by the mandibular
glands of the queen, and distributed all over her body through the
act of grooming. The molecule, as he finally identified it, had a
remarkable chemical resemblance to the ovary-inhibiting hormone of prawns. Butler, suspecting that its ovary-atrophying
effects might well work on other phyla, experimented with
various arthropods and vertebrates in hopes that this substance
might be a hormone of universal application, but sorrowfully,
he found no support for this supposition.

It would be very interesting to discover whether or not the
literal communication entity transmits an identical meaning
across phylum barriers; whether, in effect, certain words in the
chemical language mean the same thing to different classes of
animals. In some instances—we know that they do. Insect
alarm signals, the chemical exudations which comprise warning
signals, are grossly similar in classes of insects. This seems to be
generally true in animal communication. Courtship signals, for
example, whether auditory, visual (behavioral), or chemical,
or a combination, seem to be highly specific. No other bird will
respond affirmatively to a blue jay's courtship call, except another
blue jay. But almost all the birds and mammals within range of
its alarm call will respond alertly as the blue jay shrieks its warning through the woods. The same is true of insect chemical

communication, and because of this it was possible for humans to devise insect repellents (which are essentially distillations of insect chemical alarm signals) which are effective across a wide spectrum of genera. The average personal insect repellent acts upon such a disparate collection of creatures as chiggers, ticks, and mosquitoes, perhaps others as well.

Ritual food sharing among humans plays a social role very analogous to that role played by the same behavior in the societal insects. I have no intention whatever of implying that similar mechanisms are operative. It is just that I find it curious to see again and again how conservative nature is in the matter of devising systems. The same general circulatory and hormonal system is used by both plants and animals. The same flight system is employed by flying fish, bats, and birds. Jet propulsion, which was only recently adapted for marine use by us human beings, is one of the oldest marine propulsive forces extant; it is widely employed as a system throughout the mollusca (squid, cuttlefish, octopuses) and on occasion in the coelenterata (corals, sea anemones, jellyfishes), both of them ancient orders indeed.

So one can see a curious analogy between the system of sharing a food particle—and at the same time a behavior-altering chemical —employed by honeybees in the orderly regulation of their societal order, and that employed by the early Christians in the ceremony of the Holy Communion where a food particle (the wafer) is exchanged along with a behavior-altering chemical (wine) with the express purpose of forming, through the enactment of this ritual, a viable community, in Wheeler's terms, a "person" larger in scope than the collection of individual selves which comprise its separate parts.

In every human society the act of communal eating has social implications far more important than the simple act of satisfying individual hunger. Of course it is easier to prepare food for several people at once, and then only practical for these several people to eat simultaneously. But there is more to the act of social food sharing than that. In most primitive societies the sharing of food acts as an aggression suppressor. It is considered an unseemly act to attack one's dinner partner, regardless of the provocation. The fact that food sharing enhances the possibility of creating intimate relationships is well understood by business-men, fund raisers, politicians, et al. Family gatherings are far more meaningful if the gathering takes place around a dining table. All ceremonies are made more intimate by food sharing, and when a ceremony excludes the sharing of food, it is likely to be formal and the participants isolated from one another.

In many cultures, including our own, the more ceremonial the meal the more divorced it is from simple needs of nourish-ment, the more likely it will be that a behavior-altering chemical will be served with the food. This may be wine of some kind, or in cultures such as Islam, which forbid alcohol, another kind of chemical intoxicant. Even tea and coffee are stimulants, and must be considered in this category of behavior-altering chemicals. If the purpose of the ceremony is aimed at convi-viality rather than intimacy, the chemical agent is stressed rather than the food particle. But even at the modern cocktail party, where the alcohol is far more important than the food, there is still a token food offering made for the ritual sake of making a harmonious combination of these associated acts. Among tobacco smokers, smoking may be a substitute for food exchange, and the offering of a smoke is well understood in all tobacco-using cultures as a gesture of intimacy.

It is a truism among anthropologists that one can learn more about the structure of a strange society by observing who eats with whom than who sleeps with whom. On the face of it this statement should seem unreasonable, for it is obvious that the biological consequences of a carnal encounter are far more profoundly and historically far-reaching than the consequences of a shared meal. In the United States one can see superficially, just by walking the streets, from the physical appearance of most American Negroes, testimony to the frequency of carnal en-counters between the races. But those same American white men who would happily put themselves in a position to create a child with a Negro woman object strenuously to the idea of eating alongside her in a public restaurant.

It seems to me that one can only comprehend the vicious irrational antagonism to integrated dining among the civilized rationalists of twentieth-century America if one views the roots of this irrationality as reaching back into the sources and biological purposes of chemical communication as being operative in the isolation of the species, and the isolation of communities as seen from the entomological evidence. Though we have no common ancestry with insects, still the same complex of systems which propels their behavior also propels ours, albeit without our knowledge or consent.

[6]

Space

Every organism exists in space as well as in time. While the succession of forms which comprise the history of a species occurs in time, it also occurs in space, for the individual animal must possess space as well as time. It acquires time by accumulating energy through feeding, and by flight to avoid predators. It acquires space by a set of special behaviors which zoologists now call territorial.

Von Uexküll, pondering the problem of animal space perception in his attempt to formulate a set of definitions for those entities that comprised the *umwelt* of an animal, was forced to resort to Kant's definition of space which he quoted directly: "Space is merely the form of all appearances of the outward senses. . . ." We may perceive a scent, but since it is not spatial it has no form. Whatever has form for us, exists in space. Von Uexküll continued his quotation from Kant: "[space] is merely the subjective conditioning of sensibility, by which alone, intuition of the outside world is possible for us." Though Kant was speaking of humans, von Uexküll considered this a suitable working definition of space in the animal *umwelt* as well. Binocular vision, for example, which enables the two eyes of an animal to be focused on a single object, endows space with different subjective qualities than can be perceived by eyes so mounted in the head that the fields of vision do not overlap. The three-dimensional mobility of birds, fish, and certain insects endows space with subjective qualities different from the space perceived by creeping crawling things, which are restricted to motion in two dimensions. Space depends also on the total "gestalt" of the organism's perceptual apparatus—the *umwelt* space of a paramecium must be completely other in all respects from the *umwelt* space of an eagle.

Space is far more subjectively appreciated by organisms than is time, which passes for us all, great beasts and tiny animalcules alike, in roughly the same way, since the circuit of the earth, the alternation of light and dark, affects all living things on this planet, directly or indirectly. The drive of an animal to acquire time by flight is readily seen by the most casual observer. The drive to acquire space is not quite so easily or readily seen. The discovery that there is no form of life that does not manifest some behavior that could be considered territorial has only lately come to the attention of biologists. Laboratory psychologists have been concerned with the functional basis of comparative systems of space perception, but even though von Uexküll pondered the problem in the beginning of this century, the various behaviors by which the organism expressed this perception in its *natural* environment was slow in coming.

In an attempt to discover whether the basic element of space-perception, form, was perceived as such by the eyeless earthworm, von Uexküll performed an experiment based on an observation of Darwin's. "Darwin early pointed out," wrote von Uexküll, "that the earthworm drags leaves and pine needles into its narrow cave. They serve it both for protection and for food. Most leaves spread out if one tries to pull them into a narrow tube petiole [stem] foremost. On the other hand, they roll up easily and offer no resistance if seized at the tip. Pine needles on the other hand, which always fall in pairs, must be grasped at their base, not their tip, if they are to be dragged into narrow holes with ease.

"It was inferred from the earthworm's spontaneously correct handling of leaves and needles that the form of these objects which plays a decisive part in its effector world must exist as a receptor in its perceptual world.

"This assumption has been proven false. It was possible to show that identical small sticks dipped in gelatine were pulled into the earthworms' holes indiscriminately by either end. But as soon as one end was covered with powder from the tip of a dried cherry leaf, the other with powder from its base, the earthworms differentiated between the two ends of the stick exactly as they do between the tip and base of the leaf itself. Although earthworms handle leaves in keeping with their form, they are not guided by the shape, but by the taste of the leaves. This arrangement has evidently been adopted for the reason that the receptor organs of earthworms are built too simply to fashion sensory cues of shape. This example shows how nature is able to overcome difficulties which seem to us utterly insurmountable."

While von Uexküll made several curiously perceptive lunges in the general direction of territorial studies, his main fields of concern—marine animals and insects—did not provide him with much behavioral evidence to fertilize his imagination. Flight behavior, for example, may be considered negative territoriality. The possession of space is exchanged for the acquisition of time. Plants, since they are rooted, cannot exhibit flight behavior. Mobility and the possibility of flight have been sacrificed for the greater adaptive advantages of territorial possession as expressed by the extension of root systems. In the orders of botanical beings, territorial behavior reaches the ultimate height of functional development.

Just because so much territorial behavior is covert—like the root systems of plants, which extend underground and are not immediately apparent to the superficial observer—recognition by psychologists of the tremendous importance to animals of space acquisition was slow in coming.

It is not very strange, considering the importance that territorial preoccupations have assumed in shaping their national character, that it was the British who first discovered the phenomenon of animal territoriality—that is, the occupation of a patch of territory by one animal, for the purpose of excluding other members of the same species from exploiting it.

Territorial behavior becomes most obvious when it involves disputes between animals over possession of a territory. Many biologists prefer to use the terms "territory" and "territorial behavior" in this restrictive sense, as defining that geographical area which is *defended* by an animal against encroachment by others of the same sex or species. However, I shall use the term in a more general way, in its Kantian sense, to encompass the activities of an organism in response to that spatial environment in which it chooses to move. For example, the Norwegian student of barnyard fowl, Thorlief Schjelderup-Ebbe, noticed in the 1920's that status-rank in the pecking order of hens involved notions of territory, and that hen coop territories were not established for purposes of exploiting the economic potential inherent in the space—food, water, shade, or whatever—but rather for obtaining status. Many social animals, living part of their lives as members of a flock or herd or troop for whom social rank is vitally important in connection with sexual opportunity, struggle to obtain and defend territories quite independently of any economic advantages that might accrue to them. A well known example of this, which has often been filmed, occurs in the breeding grounds of sea lions in the Pribiloff Islands. The bulls come ashore some time before the cows. They establish small territories only some few yards in area, but ferocious fights take place as these territories are demarcated and taken into possession. The size of the territory that any bull succeeds in establishing determines the size of his harem which shall occupy that territory when the cows come ashore several days later.

Considering the importance that irrational territorial concerns have had in human history, it is curious how slowly recognition of the ubiquity of spatial concerns among lower forms of animal life came into being. One Francis Willughby is generally credited with being the first to comment on animal territoriality. He was a friend of John Ray, the botanical parson who was one of the most important pre-Darwinian thinkers on evolution. "It is proper," Willughby wrote in a letter to Ray in 1678, "for the nightingale at his first coming, to occupy or seize upon one place as its Freehold into which it will not admit any other nightingale but its mate." Willughby observed and reported, and that seemed to be the end of it for almost 200 years.

It was only in 1868 that another bird watcher, Bernard Altum, took over where Willughby had left off. Altum was a professor of Zoology attached to a forestry college at Eberswald, Germany. He was interested in the balance of nature, in the relationships of various forms of life to one another, in the role that insects play in the fertilization of plants, in the role that birds play in maintaining the stability of insect populations, and so on. He sought an explanation for this behavior and wrote, "It is impossible among a great many species of birds, for numerous pairs to nest close to one another, but individual pairs must settle at precisely fixed distances from each other. The reason for this necessity is the amount and kind of food they have to gather for themselves and their young, together with the methods by which they secure it. All the species of birds which have specialized diets and which, in searching for food—mostly animal matter—for themselves and their young, limit their wanderings to small areas, can not and ought not to settle close together because of the danger of starvation. They need a territory of definite size which varies according to the productivity of any given locality." Busy with his ecological concerns, and having described what he believed to be another example of Malthusian behavior, Altum occupied himself with a study of the minutiae of the dietary requirements of forest birds and overlooked the fact that he had stumbled on what was one of the most important discoveries of animal behavior ever made.

Man has always been able to identify with birds, perhaps more easily than with other animals. From St. Francis onward, bird behavior has always attracted the sympathetic human observer. Because of their ability to fly, birds have always excited the admiration and envy of men.   Also because this ability protects them from predators, they do not conduct themselves, as do so many other small animals, covertly, avoiding notice.   Birds are the only animals easily seen in great cities.   But most importantly, the *umwelt* of birds, like the *umwelt* of humans is comprised of sight and sound.   Chemical communication, scents, and odors are as relatively unimportant in the creation of avian *umwelts* as they are in the creation of advanced primate *umwelts*.   The same factors that communicate information to birds, also communicate information to us.   As a result we can perceive causal chains of stimulus-response cycles more completely from watching birds than other animals.   One need only watch the peculiarly purposeful randomness of a stray dog running in the woods to understand how much information he is continually receiving about the world via his nose, and also to realize how incomplete our knowledge of his world must be.   Scent has an inbuilt historicity.   A bloodhound can follow a scent trail several days old, while a human, deprived of an equally effective chemo-receptor, must contrive deductively from hoofprints, bent twigs, crushed grasses, etc., an intellectual reconstruction of this event that still maintains an existential reality for a dog.   Many of the seemingly random activities of small animals would become logically clear to us, were we able to join those animals in perceiving the wealth of data that they obtain through olfaction.   It is no accident that many of the pioneers of ethology discovered the basic principles of their discipline through observing birds.   Konrad Lorenz, probably the greatest living ethologist, has done his most important work with birds—jackdaws and wild geese.   Niko Tinbergen, the Oxford ethologist, came to his profession through watching the herring gulls of his native Dutch coast.   The dean of American primate studies, C. Ray Carpenter, wrote his doctoral thesis on the courtship patterns of ring doves.   The list could go on forever and would include Ernst Mayr and Julian Huxley.   Some ethologists consider Charles Otis Whitman, the first director of the Woods Hole Marine Laboratories, to be equally as impor-

tant as von Uexküll in providing the theoretical groundwork on
which the science of ethology has been built.   Whitman worked
in his spare time, as a hobby—quite apart from his professional
concerns with marine organisms—with birds; doves and pigeons.
It was as a result of the insights he obtained from watching these
birds that he was able to contrive his first tentative theoretical
formulations about the interlocking roles of genetics and behavior.

It was a series of popular books by the British ornithologist
Eliot Howard, published from 1919 to 1921, which finally
established territorial behaviors as being equally as vital an element
in the behavioral repertoire of animals as food searching, flight
from predators, and sexuality.   It was in 1933 that two students
of reptiles showed that territorial behaviors could be observed
among lizards, and in 1938 a similar study of the sex life of the
sunfish brought territoriality back into sea and marine life.
There is even some evidence that territoriality is manifested
behaviorally by such a lowly single-celled microorganism as the
paramecium.   What evidence exists is confused and highly
controversial, but it is not "consciousness" of space or spatial
relations that should concern us here, but purely a behavioral
response to space.   As early as 1909 Herbert Spencer Jennings,
the great American student of single-celled organisms, thought he
discovered in their "random" swimming, purpose and a sense of
spatial relations in paramecia.   In 1935 a German investigator,
Fritz Bramstedt, noticed that after paramecia had been confined
for an hour or so in a triangular container, then transferred by
pipette into a circular container, they continued to swim in a
triangular path.   His findings were contested by another experimenter, U. Grabowski, in 1939, so Bramstedt devised another
type of experiment.   Since paramecia have limited temperature
tolerance and tend to flee from heat, Bramstedt heated one side
of a glass microscope slide and at the same time illuminated it.
Paramecia tended to cluster on the dark, cooler side of the slide.
When Bramstedt discontinued the heat but left the light on, the
paramecia stayed on the dark side of the slide.   He claimed that he

had demonstrated learning and a sense of spatial relations through these experiments but many of his colleagues remained unconvinced. In 1952 Beatrice Gelber also claimed to have demonstrated a sense of space in paramecia, through repeatedly dipping a platinum needle baited with food into a special location in an evaporating dish containing paramecia. When she substituted a sterile needle and dipped it into the same location, paramecia clustered around it as though hoping for food. Even when no needles were dipped, paramecia tended to populate that particular area of the dish as though that portion of their spatial *umwelt* had special significance for them. Subsequent work has indicated that paramecia may involuntarily exude various chemical "markers," which help them locate themselves in space. Though involuntary, this technique is not very different from chemical territorial markers used volitionally by many higher mammals including several members of the primate order.

Considering how important territoriality had become in zoology in the 1920's and 1930's, it is surprising that it took until 1943 before territorial concepts were applied to mammalian behavior. This was finally accomplished by the distinguished American zoologist William Henry Burt, who finally wrote a paper in the *Journal of Mammology* aptly entitled "Territoriality and Home Range Concepts as Applied to Mammals." Even at this late date, however, Burt missed the principle point of the whole problem; its pertinence to human behavior and the whole science of comparative psychology. Insofar as applications went, Burt could think no further ahead than game management, and he exhorted his colleagues in zoology to study territoriality in order that animal populations in national parks not be overcrowded.

Burt's paper is still important because of a phrase—"home range"—which he had borrowed from the popular wildlife writer Ernest Thompson Seton. According to the ornithologists a territory was a defended area—an area with distinct boundaries, which an animal would defend against incursion by conspecific animals. But Burt wrote about a female chipmunk that he had observed near his house. "Although other chipmunks often invaded her territory, she invariably drove them away. Her protected area was about fifty yards in radius; beyond the fifty-yard limit around this nesting site, she was not concerned. Her foraging range, *i.e.*, home range, was considerably larger than the protected area—territory—and occasionally extended one-hundred or more yards from her nest site."

Burt used the term "home range" to describe that non-defended territorial area in which most mammals live. He said, "It seems highly probable that most mammalian females attempt to drive away intruders from the close vicinity of their nests containing young, *but this does not constitute territoriality in the sense that the term has been used by* Howard, Nice, and others *in reference to the breeding territories of birds."* Burt showed that mammalian home ranges could and did overlap, that the nest area was often defended in the same way that birds defend territories, but that the size of the foraging range depended on other considerations. Mammalian territories were different from bird territories in that the area was not necessarily closely tied to food requirements. The home range merely seemed to be that area in which the animal was comfortable, which was familiar to him. John T. Emlen, in 1948, conducted a study of the Norwegian rat in the city of Baltimore, in order to determine the home range of the rat so as to be able to understand better how it spread diseases. He discovered that rats released in their home range were immediately able to disappear into a hiding place, while rats released in territory strange to them scurried about for a long, vulnerable period looking for a place to hide. For mammals, then, territoriality had an added adaptive purpose —it consisted of that area which the animal knew well enough that when he was within it, he was relatively safe from predators. For mammals the territory is an area of such familiarity that the animal can feel secure from attack. It is not necessarily a defended area, as with birds.

The home range concept transferred from the lower orders of mammals to the hominoid level helps explain some mysterious phenomena, which have puzzled archeologists. Some sites (such as the Leakey excavations at Olorgesaile in Kenya) show signs of having been occupied for thousands of years. As layers are excavated, further layers appear below. Why should primitive peoples, not committed to settled locations by the needs of agriculture, return again and again to the same site, denuding the area of firewood and the special quartz or chert needed to manufacture their stone tools? At some of the australopithecine sites, chert was carried to the living area for a distance of twelve miles. Why? The fixation on the home range offers as good a reason as any. In modern times similar situations exist—displaced persons' camps have a habit of maintaining themselves long after their function has been served. In an alien environment this special location has become familiar, and despite its discomforts, seems secure. The present situation in Appalachia can best be explained in terms of "home range." Though many of the skilled miners of the area could find employment in other parts of the country, they prefer to remain in their home range even though they and their families exist only on a bare subsistence level. Were it not for their fixation on the home range, persons in urban slums could be more easily relocated and the slums destroyed.

Well-worn game trails are no longer effective as a refuge from predators. On the contrary, they serve as an open invitation to hunt along the known path that animals will take. How does such behavior become so firmly fixed that even under the most extreme pressure it is almost impossible to modify?

Konrad Lorenz captured some insectivore water shrews and observed their behavior when placed in a tank new to them. They used their whiskers almost as an insect might its antennae, feeling their way blindly along, whiskering each inch of a strange path. Lorenz noticed that after several inches of whiskered exploration they would dash back into the little box in which he had carried them to the laboratory from the lake where he had found them. The box served as the core of their home range. It was familiar to them, and they could only extend their appreciation of the world at large by referring themselves to it several times a minute. As he reported it, their original exploration of the dry borders of their tank was erratic. They moved slowly, with the greatest caution, whiskering their way an inch in this direction, two inches in that, and so on. By this means each of the shrews "imprinted" its way around the tank, and for the rest of their lives each of them would go from one point to another via the way it had originally "learned." After they became familiar with their routes, they could travel them with blinding speed, but forever thereafter each of them traveled the jagged path of his original exploration, none of them ever realizing what they *must* have been able to see with their eyes, that a straight line is the shortest distance between two points.

Though his shrews had perfectly adequate functioning eyes, they failed to make full use of them for either travel or hunting. Lorenz suspected they might have located their prey from a distance by visual means, but he was convinced that in the final close-in rush they were guided by agitation of the water as perceived through their sensitive whiskers. Their territorial behavior was so completely dominated by habit that Lorenz says: "One of their paths ran along the wall adjoining the wooden table opposite to that on which the nest box was situated. This table was weighted with two stones lying close to the panes of the tank, and the shrews, running along the wall, were accustomed to jump on and off the stones which lay right in their path. If I moved the stones out of the runway, placing both together in the middle of the table, the shrews would jump right up into the air in the place where the stone should have been; they came down with a jarring bump, were obviously disconcerted and started whiskering cautiously right and left, just as they behaved in an unknown environment. And then they did a most interesting thing; they went back the way they had come, carefully feeling their way until they had again got their bearings. Then, facing round again, they tried a second time with a rush and jumped and crashed down exactly as they had done a few seconds before.   Only then did they seem to realize that the first fall had not been their own fault, but was due to a change on the wonted pathway, and now they proceeded to explore the alteration, cautiously sniffing and bewhiskering the place where the stone ought to have been.   This method of going back to the start and trying again always reminds me of a small boy, who, in reciting a poem, gets stuck and begins again at an earlier verse."

As Lorenz wisely remarks, the curious defect in this behavior consists of an aberration of the memory process.   In the first instance, when the shrews leaped into the air remembering the placement of a stone that no longer existed, their unadaptive leap resulted from a defective perception.   They didn't perceive the stone's absence.   Their memory of the pathway was rigid—their perceptual system of travel, whiskering their way along by means of tactile impressions was incompatible with speed, so, like a boy reciting a poem, they learned their route not by a causal chain of meaning, but by rote.   And they tried again, and again they erred, jumping into the air inappropriately.   They failed to learn by experience.   They failed to make the *association* between their discomforting leap and the absence of the stone, which should have lain in their accustomed pathway.   Had they used their eyes as sense receptors, they would have *seen* the change in their pathway from some distance away, and not been compelled to surmount an obstacle that no longer existed.

Probably because the whole subject smacks of Lamarckian mysticism, ethology has not yet engaged itself in the question of changing *umwelts*.   An example of such a changing *umwelt* might be taken from the behavior of two types of dog in retrieving objects.   Throw a ball or a stick for a retriever, and the dog will mark the fall of the object visually, chase off in its general direction, and resort to scent location only as a last resort if the object is invisible in heavy cover.   Throw a ball out for a bloodhound, however, and the dog will chase off blindly, his nose to the ground until he can find some scent trace of the ball's bounce, then casting off into a circle, discover another trace of the ball's bounce and finally (if he's lucky) the path of the ball's terminal roll.

It is in this general area of changing *umwelts* that Lamarckian hypotheses of use and disuse may yet prove operative. Those parts of the brain that convert sight impressions into thought images develop, like the biceps of a weightlifter, through use. In animals blind from birth (and in humans deprived of sight before the age of two years) that portion of the brain—both the cortical matter at the occipital pole and the complex neural connections between the retina and the cortex—remains undeveloped.

The *willingness* of an animal to make use of its eyesight, determines the extent to which it may *imagine* sight impressions. While the eye may be perfectly functional, and under laboratory test conditions, when deprived of its other senses, the animal may display its ability to see perfectly well, it may nevertheless in its daily routine rely more faithfully on its other senses.

One of the additional reasons why mammalian territoriality remained unnoticed for so many years is that by far the majority of mammals perceive their territories by olfactory means.

Even running the risk of being boringly repetitive, I cannot resist the opportunity to stress once again the qualitative difference between the perceptual world created by chemical information and that world created by a visual imagination. That this system of territorial location by olfaction is extraordinarily adaptive is evidenced by our continued use of dogs for hunting. The fleeting, nonexistential quality of sight impressions forces us to contrive an imaginary recreation of the occurrence which does not always conform to reality. A bird hunter may hit a bird and watch its fall, marking its location as best he can in his mind's eye. But when he comes to search for its body in that spatial location he had marked in his imagination, he quite often finds that his intellectual reconstruction of the actual event is quite inaccurate. No such effort to create history is required of an animal equipped with a functioning chemoreceptor. The past exists in the present —at least as regards its chemical traces.

Domestic animals have, over the course of their domestication, lost much of their reliance on territory for subsistence and mating opportunities. Any wild animal removed from its territory and placed in another, even though the new area may be super-ficially identical, undergoes a great emotional crisis. Zoo animals transported from a cage in a zoo in one city to a cage in another may often, in their fright and panic, do themselves physical injury, or go through a long period of weight loss and apathy in their new quarters. Fully adult animals, caught in the wild and brought into captivity, often do not survive this combined shock of losing a familiar ecological territory and all the rituals of behavior that are enacted within it. For a successful life in captivity, wild animals must be captured young before their patterns of life become enmeshed with that of their familiar territory. Domestic animals can be transported around from one racetrack or show ring to another with a minimum of psychic injury. This is also true of wild circus animals, which have never been allowed to become familiar with any special territory and must adapt to being without any territory whatever from an early age. But once an animal has become accustomed to a territory, even the reduced territory of a zoo cage, it is hard indeed for it to adjust to a new territory.

Most mammals affiliate themselves with their environment by means of chemical sense-impressions—exceptions being the higher primates including man, and such marine mammals as whales, porpoises, seals, etc., and the occasional special case like the giraffe whose scent receptor is removed by height from likely contact with scent traces in its territory.

They identify their territories by scent markers, which they deposit themselves. The systems of scent markings vary with their relative usefulness to the animal concerned. For those so

constructed as to be able to travel fairly rapidly while at the same
time keeping their noses to the ground (as for instance foxes,
wolves, jackals, cats, hippopotamuses and rhinoceroses), scent
traces are deposited on the ground. The most convenient scent
traces are the feces and urine of the animal concerned. One
"housebreaks" a domestic dog or cat by defining its territory
in relation to that of its master. The interior of the house is the
master's territory, the animal's territorial orientation accom-
modates to this fact and it restricts its fecal and urinary marking
to its own territory elsewhere. Primates (including human
infants) are notoriously difficult to housebreak. Territorial
affiliation is not directly connected with olfactory marking,
though a good case can be made for the persistence of marking
habits, which among normal humans consist of optical markings
(initials, etc.). In pathological mental states, however, they may
well revert to fecal or urinary marking rituals.

The treatment of feces and urine vary as usual with the species.
Hippopotamuses and rhinoceroses distribute their feces; hippopotamuses with the special musculature of their tails, which
whirl almost like miniature propellers. The rhinoceros roots
about in his feces pile with his horn, distributing it and giving
rise to an African legend that the rhinoceros has lost his horn in his
feces and attempts to find it again—a legend that should be of
some curious interest to depth psychologists, especially in view
of the symbolic connection, which appears repeatedly in many
cultures associating the horn of the rhinoceros and human
priapean energy. For many years, because of the fame of the
rhinoceros' horn as an aphrodisiac, it was hunted hard, almost
to the point of extinction, vast prices being paid in China particularly, for powdered rhinoceros horn.

Territorial behavior of mammals is extraordinarily complex.
In addition to the special activity locations (wallows, feeding
areas, waterholes, rest areas, sleeping areas) there are activity
areas, which are used only on special occasions, during the
rutting season, or the calving season. Differing animals have

differing rituals connected with these territorial locations. Ethological studies of the interlocking relationships between activities and territorial locations are just beginning.

We still possess, in the structure of our shoulder bones and muscles, evidence that our ancestors were arboreal. The baboon, a terrestrial monkey, cannot hang from his arms as we can.   Anthropologists surmise that our human ancestors, possibly several species of apes rather than merely one apish ancestor, descended from the trees at some later time than did the ancestors of the modern baboons.   Fecal marking is not very useful to arboreal creatures; some of the lemurs mark with urine, others with special scent glands usually located under the tail.   But in the higher reaches of the primate order, territories are established by visual recognition, a form almost unique among mammals.

That traces of territorial marking instincts still persist among humans can be established by a wealth of detail.   For example the "Kilroy Was Here" drawings of World War II are undoubtedly territorial markings.   The keepers of public monuments fight a losing battle against the scrawled, carved, scratched legends that visitors leave.   But perhaps the most directly territorial marking by humans is the urinal graffiti.   As with animals, it is the male of the human species who is the most ardent marker; and the compulsion to mark insulting legends on the walls of urinals seemingly transects all economic classes and educational levels.   We have all seen graffiti on walls where we should never expect to see them, in universities and exclusive clubs.   Mating area markings are a commonplace among animals.   Among tigers the female leaves a special urine trace on the male's marking spot as a sexual invitation, and on their next encounter he is prepared for her presentation.   Among humans the park benches and tree trunks adjoining a popular assignation area are overlaid with a mosaic of carved initials often joined with Cupid's arrow and enclosed within the outline of a heart.

Among humans the association between the related ideas of

geographic and personal possession is demonstrated by the word *property*, which includes both real estate and personal belongings. For young American males of mating age the first territorial possession is not a geographic area but an automobile, the sexual significance of which is well understood by manufacturers and their advertising agents.   In this respect the behavior of the American male is not very different from that of many birds and mammals.   The acquisition of a territory is an indispensable commencement for any courtship activity.   The American male must often get a car before he can get a mate.

Territorial marking of personal possession is also a compulsion with most small boys who make their "special mark" before they can write on all their precious belongings.   Though they may claim such markings are for identification purposes, it is usually the most unique possession that is so marked, the very possession that if lost and then found again could be most easily identified without being specially marked.

According to Freudian doctrine, the anally fixated personality is characterized by stinginess, penury, an obsessive concern with personal property, all of which is associated in this neurotic pattern with an equally obsessive concern with excretion and feces.   Freud considered that this correlation of concerns had causal connections; that as a result of early childhood experiences connected with toilet training, the anal neurotic made an unconscious association between feces and possessions.   According to the animal behavior patterns noted by Hediger and other ethologists, property and feces are connected by utility—property is *demarcated* by feces markings.   In some species (foxes and civet cats) anal scent glands distinguish the scent of the feces as they pass through the anal orifice; in still other animals (skunk, polecat) these anal scent-marking glands have become even more evolved and specialized to the point where they are now part of the defensive armament of the animal.

Hediger notes that among the Berber tribes of North Africa, olfactory traces are still used as territorial markers; special

evil-smelling concoctions are sold in the bazaars to be used by
house-holders, who hang these dreadful talismans from the eaves
to ward off evil demons.

It must be obvious at this point that there is more to territorial
behavior than the simplistic economic exploitation of the territory
that Altum described.   As structures evolve, they often lose their
original function and take on another: the anal scent glands of the
skunk are an example.   A similar alteration of function occurs
as behavior evolves, so while Altum was doubtless correct in
assuming that bird territorial behavior was originally adaptive for
the direct exploitation of the economic potential of the feeding
and nesting area, the behavior has, in the process of its evolution,
become increasingly complex.

In many social animals and certainly in man, territorial acquisition has become associated with social rank.   Competition,
between individuals who are members of a group, for any opportunities—for access to food, or sex, or space—involves the
competitors (as they persist in competing over a period of time)
in a relationship.   After a series of competitive encounters one
animal will tend to dominate the other.   By this means, social
rank is established.   The higher ranking animal in a group will
have greater access to all the commodities available than the
lower ranking animal.   By establishing such a social rank order,
the group can cohere with its members in close proximity to one
another.   Animals incapable of recognizing one another's social
rank exist in groups that are widely dispersed.   This is well
described by C. Ray Carpenter's report of the social life of the
gibbon.   Almost alone among primate species, the gibbon lives
apparently peacefully and monogamously in widely dispersed
large groups.   At first this was thought to be due to the peaceful
disposition of the animals, but Carpenter showed that the animals
were in fact so highly competitive and mutually hostile to one
another that they lived dispersed from one another because of
what he called "hostility spacing."   The effect of this dispersion
was the peaceful conduct of the troop.   Because of its totally

arboreal way of life the gibbon can survive relatively independently of other members of the troop. There are few predators dangerous to him from whom he cannot escape by swinging through the trees like a trapeze artist.

The baboon, however, living on the ground, is almost totally dependent on other members of the troop for mutual support. A single baboon is no match for a leopard, but a single leopard is no match for three or four ferociously armed, determined baboons, each attacking him from a different quarter. Because their terrestrial way of life demands it, the baboon troop must perforce travel quite cohesively, each member easily accessible to another in case of emergency. Living, as they do, in close quarters, each member of the troop must play a specialized social role depending on sex and age. The social roles of males and females are completely different. As the troop moves out on the march, the females and their infants form a kind of ragged hollow ring with the young adult males on the outside of this ring. The social role of the so-called subadult-adolescent-male is different from that of the fully mature male. These subadults form special "uni-sexual" bands and are only fully integrated into the troop proper, when they are fully mature. It is here, in the uni-sexual bands, that many of the male status quarrels are settled early in their adult lives. These young males are also the expendables— they straggle along the line of march at the outermost periphery of the troop, usually in front of the main body, or in the rear. These are the first animals who would normally encounter a predator; they would send up the alarm and engage the predator until the adult males could arrive in force from within the main body of the troop and deal with it.

Seen casually, without any attempt to impose an order upon it, the baboon troop on the march looks like a ragged assembly. But after one has seen a sufficient number of these assemblies, they take on a pattern. At the point in the direction of movement are the young males mixed with subadults. Behind this point straggles the main body of the troop, males of various ages with no particular status ranking, females with young, and aged of both sexes. In the center of this assembly and walking with marked arrogance, almost a strut, the tail held up stiffly from the root for one-third of its length in the typical baboon manner (which, to the human observer—accustomed to seeing ungulate tails drooping gracefully from the root, obeying the force of gravity—seems to be a deformity) may be found the members of the so-called dominance hierarchy. These are the status-rich males, the acknowledged leaders of the troop. They usually move at a leisurely pace, somewhat separated from one another. After seeing a number of baboon troops on the move, one is struck by the fact that the centrality of these large and handsome adult males seems to be isolated from the rest of the troop. Baboons tend to move slowly with much stopping and examining of possible food particles by various members as they march. The dominance hierarchy males also stop and sit quite frequently with their heads quite often held high, the chins raised in that posture stereotyped by the Egyptian artists representing Thoth.

Studies by various teams of baboon-watchers, particularly the American anthropologists Sherwood Washburn and Irvin DeVore and the zoologist Stuart Altmann, have demonstrated that this isolation of the dominant males derives from a curious retraction of territoriality, which DeVore calls "personal space." For the baboon, personal space is established by the act of seating himself. By the very act of ceasing to walk forward, by actually sitting down on his haunches, the dominant male establishes, around his seated person, a territory, in the ornithological sense, a private zone of space into which another member of the troop may only intrude at the peril of attack. This personal space represents a territory within a territory; for the troop as a whole has a collective territory, in which it forages and travels in relation to other troops of the same species occupying the same general area.

Personal space becomes apparent whenever the troop stops its march and the animals seat themselves. If a socially inferior male must, for some reason such as to avoid a natural impediment (a boulder or tree), invade the personal space of his superior, he does so with unmistakable signs of servility, with a cringing sideways, tentative approach, and by grimacing. The baboon grimace is a horrid, sterile smile, a baring of the teeth in malevolent caricature of the human smile. The smile of a chimpanzee or a gorilla can indicate amusement like the human smile, but the grimace of a baboon seems to be either a gesture of ingratiation, or menace.

It is very likely that the establishment of personal space on the part of dominant social animals, whether baboons or seagulls, serves an adaptive purpose. For birds and primates, most of the communication message is transmitted by gestural movements. The distinctive calls, cries, coughs, grunts, etc., merely serve to draw the glances of others to the communicator. Thus, being surrounded by personal space the dominant animal is immediately and distinctly visible to all the members of his society who are then in a position to respond appropriately to his signals.

There is rarely an attempt on the part of low-ranking individuals to establish personal space (except for an occasional maverick or an ill animal). In fact, the low-ranking members seem to draw satisfaction from the immediate proximity of their fellows. This phenomenon of personal space has been noted in many species of social animals other than baboons. In his wonderful book *The Life of the Herring Gull*, Niko Tinbergen shows a photograph of sea gulls spaced regularly along a breakwater, each separated from the other by a space appropriate to his social rank. This desire to acquire personal territory usually appears in adulthood; young animals prefer to annihilate personal space by swarming over one another in the nest or litter. This altering need for territorial space is evident among humans as well—children liking close personal contact and dominant adults preferring ever greater degrees of solitude. The private office, the secluded country estate, are prerogatives of high social rank in human society. Conversely, many urban redevelopment projects have caused their inhabitants discontent by providing too much personal space for those who, by reason of low social rank, gain a feeling of personal security from the close spatial relations with others, which prevailed in their crowded dwellings.

Dominant animals acquire their personal space by what ethologists call "threat behavior." Often, among the lower vertebrates, rituals of threat are combined with special bodily organs, which have evolved as communication systems. An example from the American chameleon, as described by L. T. Evans in 1936: "The urge to acquire and to hold territories against other males is very marked. The resident male, upon the approach of a strange male, will extend a dewlap or fan lying in the mid-ventral line beneath the lower jaw. This is a challenge or warning to the stranger. If the latter does not reply to this challenge, no fighting is apt to occur although the resident may persist in his challenging display for some minutes. If the nonresident also flourishes his dewlap, then the dorsal crest along the neck of each male rises slowly to a height of perhaps four millimeters, and they begin a sidewise approach toward each other while they flatten their sides to such an extent that the belly drags along the ground. When they are within six inches of each other, they continue to strut back and forth with dewlap flashing and body flattened. Many encounters end at this point. The male that is most impressed with the display of the other moves off, lowering his crest and withdrawing his dewlap, his body no longer flattened."

Among the mammals, and particularly among the primates, many threat signals, like warning signals, appear to be universally understood between the species. Humans understand, for example, the threat display of a dog, the baring of the teeth, the rising of the pelt along the back of the neck, the growling, and most of all, the threat contained in the eye glance. The dewlap threat display of the lizard, on the other hand, is not necessarily threatening—no more than the blooming throat of a bullfrog in mid-song. Among all the primates, the long, level, baleful look usually comprises the beginning of a threat display. We can easily read the meaning in this glance whether it is delivered by an ape or by another human. It is a frightfully subtle kind of ethological understanding, defying quantitative or qualitative description: It's not a matter of how long the stare is maintained, nor of describing the actions of facial muscles around the eyes, it is a communication delivered to us at a level so deep—long predating speech—that we know it all too well to be able to discuss it "scientifically."

A dominant animal is somehow able to acquire personal space by displaying these threats, while the subdominant animal cannot. He takes what space he can by moving off to the periphery of the group when he can; and when he cannot, under conditions of confinement, he suffers. John J. Christian, of the Johns Hopkins School of Hygiene, and David E. Davis of Pennsylvania State University weighed the adrenal glands of mice after social rank had been established in a confined space. In this situation—where the animals were caged, making escape impossible—actual fighting took place. Mere threat behavior could not provide sufficient space. Christian and Davis found that the adrenal glands of low-ranking mice were enlarged, the cortex inflamed even though they fought less and had fewer social interactions than higher ranking mice whose adrenal glands were normal in appearance and less enlarged.

There comes a time, even in low-ranking animals, when close spatial relations stop becoming socially supportive and become emotionally destructive. The emotional stress produces physiological damage. Arteriosclerosis, a common ailment of urbanized humans, is one of the consequences of territorial deprivation. In 1958 Herbert L. Ratcliffe and Michael T. Cronin, knowing that arteriosclerosis is a common cause of death among zoo animals, reviewed the autopsy records of mammals and birds that died in the Philadelphia Zoological Gardens over a period of forty years from 1916 to 1956. This covered 3,360 mammalian deaths, and some 7,600 avian deaths. It had been thought that fatty diets contributed to the formation of cholesterol which produces arteriosclerosis. Ratcliffe and Cronin chose this particular period of the Philadelphia Zoo records because in 1935 a new, high-protein, low-fat diet had been introduced and they were interested to see what effect, if any, it had on the rate of arteriosclerosis among zoo animals. They were most surprised to discover that instead of reducing the number of deaths due to arteriosclerosis the new diet increased the incidence of the disease! Among mammals there was a ten-fold increase, and among birds a twenty-fold increase. There was also a marked increase in the number of deaths due to combat injuries, which suggested to them that territorial quarrels had become more acute, especially since on examining the records of their confinement they found that, among females, lactation had been suppressed, ovulation cycles disrupted; among the males autopsied, the manufacture of sperm had been suppressed, and in both sexes enlarged adrenal and thyroid glands were discovered.

Since territorial aggressiveness is more of a behavioral constant from species to species among birds than it is among mammals, the dramatic increase of arteriosclerosis among the avian members of the zoo population suggested to Ratcliffe and Cronin that territorial deprivation had something to do with the disease. There had been some increase in the population density of the zoo during the forty-year period, but many animals had continued to be exhibited in separate cages, so territorial interactions could not be the only explanation. Ratcliffe and Cronin finally suggested that the new diet adopted in 1935 gave the animals more energy, thus expanding their territorial ambitions. As they were frustrated in these ambitions, the classic side-effects of overcrowding—pathologies in the reproductive system (ovulation, lactation, the manufacture of sperm in the testes)—began to be noticed. And arteriosclerosis!

Man has long known that many animal populations undergo cycles of increase followed by a sudden "crash" or die-off. The nomadic way of life of primitive hunter-gatherer peoples was adopted to take maximum advantage of these cycles. When game became scarce in one area, the tribe would move to another. Today game managers attempt to control the cycle to some extent by extending hunting seasons, culling herds, permitting the shooting of females, and various other devices.

The phenomenon of population cycles has fascinated man from the beginning of time and there is a vast literature on the subject, running all the way from Pharaoh's Biblical dream of the seven fat kine and the seven lean to more modern, sophisticated attempts to understanding it. Cycles among mammalian populations are of most interest to man at this point of time when his himself faces, from within his own species, a population explosion of unprecedented magnitude. The human problem is overwhelmingly complex in its interior detail, but if it is viewed within the frame of the animal evidence it can be seen in large part to be basically a problem of territoriality. The animal evidence demonstrates quite convincingly that the population "crash," when it occurs, is not due primarily either to starvation or disease as was heretofore believed, but to the psychic stresses of overcrowding which, among animals, have immediate and usually lethal physiological consequences.

In 1950 John J. Christian, the Johns Hopkins endocrinologist, wrote a paper analyzing the field studies of population crashes—mostly of rodents where the phenomenon occurs quite regularly and is easily predicted and studied. One curious facet that he discussed had to do with the fact that even when animals involved in a population "crash" were trapped, removed from the frightful pressures of their natural environment, and cared for in captivity, the physiological processes that were triggered by territorial stress proceeded to their deadly conclusion. Christian writes: "During an intensive study of the periodic die-off in Minnesota [biologists] were able to demonstrate that a very small number of deaths could be attributed to infectious disease. The majority of animals exhibited a characteristic syndrome which the authors have termed 'shock disease.' This syndrome was characterized primarily by fatty degeneration and atrophy of the liver with a coincident striking decrease in liver glycogen[1] and hypoglycemia[2] preceding death. Petechial or ecchymotic brain hemorrhages and congestion, and hemorrhage of the adrenals, thyroid, and kidneys were frequent findings in a smaller number of animals. The hares characteristically died in convulsive seizures with sudden onset, running movements, hind-leg extension, retraction of the head and neck, and sudden leaps with clonic seizures upon alighting. Other animals were typically lethargic or comatose." Referring to Dr. Hans Seyle's classic definition of shock, Christian demonstrated that the loss of blood sugar had nothing whatever to do with dietary intake, but was rather caused by malfunctioning of the secretions of the pituitary gland which in turn affected the adrenal glands. Seyle termed the syndrome the "exhaustion phase" of countershock. It could be produced by a number of means including exposure to cold, injections of formaldehyde, or intense, continuous emotional stress. The investigators of the snowshoe rabbit die-off "showed that the convulsions in hares did not occur until the liver glycogen dropped below .2 per cent, demonstrating that they were caused by a progressive fall in the glucose reserves. Experimentally, they were able to stop or alleviate the convulsions temporarily by intravenous glucose injections. . . . Epinephrine injections were ineffective. It seems more than likely that in most die-offs in mammals, we are dealing with a manifestation of [Selye's] adaptation syndrome, with the terminal convulsions precipitated in many cases by some sudden stress, such as fear or exertion."

In 1960 another investigator, John B. Calhoun, wrote a paper describing how overpopulation can occur *with the consent of the animals involved*, and showing how behavior gradually becomes pathological, producing those emotional stresses which in turn produce a physiological backlash. Prior to Calhoun's experi-

---

[1] Animal starch, a source of energy produced by the liver.
[2] Decreased amount of sugar in the blood.

ment there had been numerous studies of overcrowding and its stresses, but as one reads Calhoun's paper, one is struck by his own personal consternation and incredulity as he watched this ferocious drama make itself manifest. In his experiment over-crowding took place even though he had provided the animals with sufficient space to disperse if they had wanted! Sometimes animals cannot disperse themselves because of geographical barriers, *i.e.*, island, rivers, mountain ranges, etc., but often there are no geographical barriers—only psychological ones.

Calhoun has spent a good deal of his life as a psychologist studying the wild Norway rat, that frightening beast that has, in the last two centuries, traveled in man's company throughout the inhabited world, spreading disease and havoc wherever it has gone. Contrary to what its name would suggest, biologists now believe the Norway rat originated in China, and only began to appear in Europe during the latter part of the seventeenth century. It has defied man's most ingenious attempts to exterminate it. Calhoun has devoted much time and effort to an attempt to understand its habits, motivations, and the social organization of rat communities. He began to suspect that rat social structure was dependent to some extent upon the way in which access into the communal nesting burrow was obtained. He contrived an experiment by designing two different types of nesting burrows, each similar to those rats inhabit in the wild—one with almost unlimited access, and the other with only four entrances. Each burrow would house twenty rats, there would be four burrows in a room and four roomsful of burrows. To his amazement and dismay he discovered quickly that his experiment was going awry, that something was terribly wrong. It had nothing to do with the way in which the burrows were constructed but the way in which he had placed the four burrows in each room.

Each room had been designed like a one-story apartment house. The four burrows were separated from one another by grids of wire mesh. Rats could pass through holes in the mesh from one enclosure to another, but the enclosures were lined up like an old fashioned railroad flat.   To go from enclosure 1 to enclosure 4, a rat would have to pass through enclosures 2 and 3.   This proved to be the critical factor; gradually, enclosures 2 and 3 became the focus of social activity.   Rats from enclosure 1 would visit with those in enclosure 2, those in 2 would visit with those in 3.   But the rats in 4 would also visit with those in 3, so that over the course of two years and three generations of rats, enclosures 2 and 3 became overcrowded, housing just double the population density that Calhoun had hoped for, while enclosures 1 and 4 were almost deserted.   This took place (with some minor variation as to whether enclosure 2 or 3 became the more horrendous) in each of the four rooms that Calhoun had constructed. He had designed each room to support a population of eighty rats, and he never permitted the population to rise above this level.   During most of the course of the study, the population was well under this figure; there were only forty-eight rats in each room by the end of the second generation; and already trouble was well underway.   By the time the third generation had come of age and the population had reached its peak of eighty rats in each room, complete chaos came to pass—a horrid reminder of what conduct in the medieval and early industrial age slums of Europe must have been like, and what much human slum life is like today.

Each pen had its own supply of food, water, and paper strips for the females to make nests out of at litter time.   The dominant males who asserted their territorial privileges, naturally took control of the feeding stations.   Calhoun had anticipated this and provided for it by making the hoppers large enough for all to feed without actual combat.   There might be threatening if a subdominant came too close to a dominant, but the former could move away and still be able to eat and drink.   However, rats are convivial creatures; they like to eat in one another's company. Seeing one rat eating seems to remind the other that he is hungry and also wants to eat.

Calhoun had begun the experiment by placing pregnant rats in each enclosure and starting his experiment with animals that were litter mates, and therefore of the same age.   One of the curious and unexpected things Calhoun discovered was that the dominant territorial animals appeared to mature later than the subdominants.   By the time the dominant rats in enclosures 1 and 4 had begun to assert their dominance, there was no one to dominate.   The subdominant animals had already begun taking their meals with their friends in the neighboring enclosures. The nasty, quarrelsome dispositions of the dominant males, ready for a fight but frustrated by being unable to find anyone to fight with, accelerated the flight of subdominant animals from the end enclosures.   After a time there was no sense in coming "home" even to sleep if, waiting at the hole in the mesh, was a vicious, aggressive, territorial male ready and waiting for a fight.   One of the functions of dominance in animal societies is to regulate and stabilize group living.   Once the hierarchy is established, everyone gets along fairly well.   One of the factors causing chaos in this experiment was the formation of a biological community without the evolution of a social community due to the fact that dominant territorial males failed to emerge before the group dispersed.

Invariably, social dominance is combined with territorial aggressiveness.   As the feedhoppers in the center enclosure became more crowded, threat behavior no longer sufficed to enforce the rights of rank, and actual fighting began to take place.   Also, dominant males acquired harems of females at the expense of subdominant males.   This led to further fighting, and to sexual aberration.   Calhoun writes: "The first sign of disruption involved more frequent attention to and attempt at mounting females who indicated no sign of being receptive.   Later males mounted other males, and a few of these, particularly third [generation] males seemed to prefer other males as sexual partners. In the final phase, young rats, even recently weaned ones of both sexes, were mounted."   The normal rituals of rat courtship involve what Calhoun calls the "pursuit phase."   As he describes it: "Following the withdrawal of a receptive female into a burrow, the male follows her to the burrow opening, but does not pursue her into the burrow. He waits there quietly or with intermittent movements back and forth with his head protruding into the opening. The full sexual dance of wild Norway rats did not develop fully in the artificial burrows in the absence of the conical mound of dirt. Eventually the fully receptive female emerges and is pursued by the male until he over-takes her. The chase culminates in intromission as the male mounts the receptive female while holding her neck with his teeth so gently as not to cut the skin. Simultaneously he exhibits pelvic thrusting as she exhibits lordosis."

However, as overcrowding became apparent, there developed what Calhoun called the "behavioral sink." He notes that "pursuit also became altered. With increasing frequency, males who followed a receptive female to the burrow also followed her into and through the burrow. Such intrusion produced further disturbance to lactating females (who were already inside the burrow). . . . Another element of disruption of normal sexual behavior involved the scruff-of-the-neck biting act during mounting. As the behavioral sink became accentuated in its influence, many females following their period of receptivity were characterized by literally dozens of nicks about the dorsal aspect of the neck. Males subjected to homosexual advances exhibited similar wounds but fewer in number . . . abnormal aggressive acts developed as the behavioral sink became established.

"[One] of these was tail biting. A peculiarity of this behavior was that males alone exhibited tail biting insofar as I could deter-mine. Furthermore the population became divided into tail biters and those who were bitten on the tail. The latter category included both sexes. At times it was impossible to enter a room without observing fresh blood spattered about the room from tail wounds. A rat exhibiting tail biting would frequently just walk up to another and clamp down on its tail. The biting rat would not loosen its grasp until the bitten rat had pulled loose. This frequently resulted in major breaks or actual severance of the tail. . . . The basis of this behavior has so far eluded me. From several very incomplete lines of evidence presently available I can only say that I suspect that tail-biting derives from a displacement of eating behavior rather than stemming from a modification of aggressive behavior. . . . [Another] aberration takes the form of slashing attacks. Gashes ranging from ten to thirty millimeters may be received by either sex on any portion of the body. The depth of such wounds frequently extend down into the muscles or through the abdominal wall."

The endocrine disturbances that result from overcrowding are well known and almost always involve difficulties in reproduction. Ratcliffe and Cronin alluded to it in their report on Philadelphia Zoo animal autopsies. But let Calhoun tell it in his words: "There was a reduction in conception, or at least a reduction in pregnancies continued to the age when embryos could be detected by palpation. Also, pregnant females exhibited difficulty in continuing pregnancy to term, or in delivering full-term young. Both phenomena were noted only after the behavioral sink began developing. Several females were found near term lying on the floor with dark bloody fluid exuding from the vagina. I never found any evidence that these females delivered. One died while I watched her from the overhead observation window. She was immediately autopsied. Extensive dark hemorrhagic areas in the uterus suggested that the fetuses had died before their mother. Another apparently full-term female was autopsied shortly after death and found to contain several partially resorbed full-term embryos.

"Upon palpation for pregnancy, more and more females were recorded as containing large hard masses in the abdomen. These sometimes reached a diameter of ninety millimeters. Usually death occurred before attainment of such size. A group of females in which these abdominal masses existed were autopsied. The enlargements proved to be thick-walled dilations of the uterus. Usually these dilations contained a purulent mass. Partially decomposed fetuses were found in some of the rats in which these dilations were still relatively small.

"Eleven females with these masses, some from each of the three [generations] were autopsied by Dr. Katherine Snell of the National Cancer Institute. . . . Some of these eleven rats as well as four others autopsied because of obvious mammary tumors showed the following pathologic lesions in one or more females: (1) fibromyoma[1] of a uterine horn; (2) fibrosarcoma[2] of the mammary gland; (3) fibroadenoma[3] of the mammary gland; (4) angiomatous[4] adenoma of the adrenal cortex; (5) granulomas[5] of the liver; (6) papillary[6] cyst of the thyroid." However Calhoun is quick to write: "No conclusion is warranted concerning the influence of behavior upon incidence of the tumors." He continues that the sample was highly selected, that ink used to mark these rats contains substances that are "known to be absorbed through the skin, to be toxic, and some possibly carcinogenic." Yet he admits that other control rats marked with the same substances did not experience reproductive difficulties, or develop tumors.

The rats experienced urination difficulties, hemorrhaging bladders, blocked urethras, as well as bladder and kidney stones. This suggested to him that there was some malfunctioning of the vitamin A metabolism. Once again, after the vitamin A content of the livers of a control group had been analyzed and the vitamin content of the food assayed, no dietary explanation for the pathologies could be found.

One last point—in human slums, children are well known to be ill-cared-for and neglected. There have been numerous sociological explanations for this, usually connected with the fact that low-income group mothers must work to supplement the family funds. Calhoun's data suggests that in addition to this obvious factor, there may be still another additional force at

---

[1] A nonmalignant tumor composed largely of fibrous tissue.
[2] A malignant growth approaching normal fibrous tissue in character.
[3] A benign tumor of glandular origin approaching fibrous tissue in character.
[4] A tumor composed chiefly of dilated blood or lymph vessels.
[5] A morbid growth consisting largely of granulation tissue.
[6] Nipple-like.

work—an impairment of the normal maternal protective "instinct." "When a female while with her young is disturbed," writes Calhoun, "either by the experimenter, or by an invading strange rat, she customarily transports the entire litter from one place to another. Such a transport is not interrupted by any other behavior until the entire litter has been moved to the same place. . . . In the normal condition when females had litters in separate boxes the litters were maintained intact with no mixing. As the behavioral sink developed, litters became more and more mixed. When only one litter was present in a burrow, the young frequently became scattered among several boxes. This resulted from the female's interrupting the transport behavior by some other behavior. The consequence of a reinitiation of transport was likely to be some nesting other than the first. In the extreme state of disruption the terminus of transport was undirected. The mothers would take a pup out of the burrow and start toward the floor with it. Anywhere along the way, or any place on the floor, the mother would drop the pup. Such pups were rarely if ever retrieved. They eventually died where dropped and then were eaten by other rats."

Normally, rats build neat and comfortable nests for their young. Calhoun writes: "Failure to organize paper strips taken into nesting boxes formed the first indicator of disruption of this behavior. Although many strips were transported they were just left in a pile and trampled into a flat pad with little sign of cup formation. Then fewer and fewer strips reached the nesting box. Frequently a rat would take a single strip, and somewhere along the way it would drop the strip and then engage in some other behavior. In the extreme state of disruption, characterizing at least all major pens of feeding, the nesting material would remain in the center of the room for days. Even when females delivered litters in the burrow of the major pen of eating, no nest was formed; the young were merely left on the bare sawdust periodically placed in every box by the experimenter."

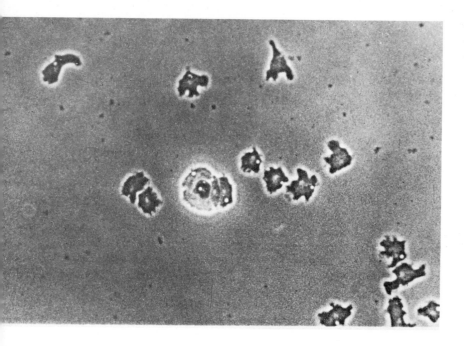

Taken by Dr. G. Gerisch of the Zoological Institute in Freiburg, this photograph shows the rounded founder cell seeming almost to glow in the center of the picture, attracting to itself other social amoeba during the preliminary stages of aggregation. On the right hand side of the founder may be seen another animal beginning to climb atop the founder, forming a clump (*Courtesy of Dr. G. Gerisch*)

(From GERISCH, Roux' Arch. Entwickl.-Mech. Org. 155, 342-357 (1964). Springer-Verlag, Berlin · Göttingen · Heidelberg.)

In this picture, taken by Dr. Kenneth Raper, the streams of animals moving toward the central collection points may be clearly seen (*Courtesy of Dr. Kenneth B. Raper*)

PART THREE

Social Organization

[7]

Sociality

There exist among the rotting leaves on forest floors, a class of creatures known technically as the *Acrasiales* (from the Greek, meaning *unmixed*) and popularly as the slime molds.

As they perform their ecological function of contributing to the decay sequence they live as amoeba, freely, each wandering separately over its chosen area ingesting bacteria by *phagocytosis*, by surrounding and enclosing its prey animal, incorporating it literally into its flesh.   The following description of its behavior, while generally true of all the eight or nine species of this class, refers particularly to the species *Dictyostelium disocideum*, which was identified in 1935 by a soil chemist named Kenneth B. Raper, who worked at that time for the United States Department of Agriculture in Washington.   Raper found that this particular species of slime mold (unlike many others) submitted easily to conditions of captivity, flourishing under a wide range of humidity and temperature gradients.   Because of this it soon became the primary animal used by laboratory students of its behavior.

This behavior is unique, and as a result, fascinating to a variety of biological disciplines—to cytologists, embryologists, geneticists, and others.   It should be the object of study for sociologists as well, since this tiny brainless, nerveless animal has evolved a complex system of social interactions; yet for some strange reason it has remained neglected.   Perhaps it is because animal sociology still remains in its infancy, and seems to be progressing backwards down the ladder of biological complexity from the more easily observed and presumably more easily comprehended social behavior of vertebrate forms to the social behavior of more simply constructed beings.

While in their free-living state, each amoeba lives alone, moving about along the substrata either by elongating and then contracting its entire body, or else by extending and then contracting extruded portions of itself called *pseudopodia*. Its form is always in flux.

When food is plentiful and other conditions appropriate, every amoeba multiplies by simple cell division once every three or four hours. Eventually this geometrically expanding population exhausts the available local supply of bacteria, and as this happens the amoeba commence the enactment of an incredible series of activities. These activities are a literal metaphor for the organization of cells in a multicelled individual, or the organization of individuals into a social unit, whether that unit be an ant colony, a baboon troop, or a human society.

As their food gathering becomes more difficult, the wandering amoeba begin as individuals to cease their feeding and begin to form communal aggregations: First a few individuals clustering around a dominant (or attractant) individual, and then this group joining other groups until (as seen on an agar dish) clumps of organisms discernible to the naked eye form themselves, giving the surface of the dish a stippled effect. Now the clumps begin still another aggregation—they begin to join one another, clump to clump. They form straggling streamers of living matter, which now begin to orient themselves toward central collection points. At this stage the dish seems to be covered with numerous, regularly spaced, many-armed swastika-shaped patches of slimy mold. At the hub of each central aggregation point, a mound begins to form as groups of amoeba mount themselves atop other groups, which have already arrived at the central hub. This hub gradually rises first into the shape of a blunt peg, and then into a distinctly phallic erection. When all the incoming streams of amoeba are almost completely incorporated into this erected cartridge-like form, it topples over onto its side, now looking like small, two-millimeters-long, slimy sausage. This slug begins now to migrate across the forest floor to a point where, hopefully, more favorable ecological conditions will prevail. At this point the communal mass that forms this slug is known as a *migrating pseudoplasmodium*. It seems to possess a discrete envelope, almost a skin; but this is a sheath of slime, and as it migrates it leaves behind a trail of slime. It looks exactly like a minute garden slug, except that it lacks the extensible horns of these snails.

As the slug migrates, it continues to attract scattered solitary amoeba, which had not participated in the original aggregation. These join the mass and become immediately incorporated within it. Estimations about the size of the population that comprises the average slug vary, but generally it is thought that perhaps some half a million amoeba are involved.

When he discovered this animal thrived in the laboratory, Raper experimented with directing the course of the slug's migration and found that it responded to light and warmth; such little light, as a matter of fact, as that provided by the luminescent dial of a wristwatch shining in the dark, and such little heat as 5/10,000 of a centigrade degree. He found he could lead the slug around over a smooth surface by the light of a watchface just as one can lead a donkey with a carrot on a pole. Raper also noted that the slug narrowed quite obviously into a point at its front end. "During migration," he writes, "the point, the apical tip, as it has been termed, is constantly to the fore and apparently guides the migration of the entire body." Raper cut this tip off the slug and found that when the community was thus deprived of its leadership, migration stopped dead in its tracks. "When the anterior portion of a migrating plasmodium is removed," he writes, "the decapitated body ceases migration, nor does it respond to light. The amoeba comprising it, crowd forward to the line where the excision was made, and there collect in a rounded body. . . . In only a few isolated cases has a pseudoplasmodium thus decapitated been observed to form a new apical tip with accompanying directive center."

After migrating for a variable period of time (which can be two minutes or two weeks) in the direction of light and warmth, this slug, this wandering community, now ceases its movement and enters into another phase of its communal history called "culmination" or "the formation of the fruiting body."   Exactly what causes migration to stop is not yet known, but as it ceases its forward motion, the slug gradually erects itself once again into its phallic shape until it is standing on its tail.   John Tyler Bonner, a zoologist at Princeton who has spent the major part of his working life studying these fascinating creatures, describes what happens: "In the apical tip of the pseudoplasmodium, a group of cells near the tip becomes rounded off and enlarged.   The whole group is either round or oval in shape, and its outer limit is smooth, bounded by a visible wall. . . ."   This oval shape gradually assumes the form of a candle flame, bellied at the bottom and coming to a point at the top.   As the belly forms, a waist also appears between the base, the tail of the now-erect slug, and the candleflame section which is forming directly above it.   This waist gradually lengthens and becomes a stalk, pushing the candleflame section ever upwards into the air.   This brittle stalk continues to rise, carrying the candleflame section which now tends to become more spherical in shape, up and up, either in a straight line, or in a wavering direction, depending on certain conditions.   Normally the end of this culminating stage produces a brittle form which looks exactly like an old-fashioned hatpin except that it is smaller—only half an inch or so high.   The candleflame pod atop the stalk is known as the spore mass.   As this process unfolds over the course of some two or three hours, it seems as though the stalk, growing from the bottom upward, lifts the sorocarp or fruiting body into the air.   But what actually happens (as Bonner demonstrated by staining certain cells and noting the direction of their movement) is this: Amoeba from the apical tip migrate downward through the forming fruiting body,

thus erecting the stalk.   As the process actually occurs, it resembles
a slow motion movie of a rising water fountain run backwards.
Bonner describes it somewhat differently: ". . . to accomplish
this transformation, the slug first points its tip upward and stands
on its end.   The uppermost front cells swell with water like a bit
of froth and become encased in a cellulose cylinder which is to
form the stalk.   As new front cells arrive at the tip of the stalk,
they add themselves to the lengthening structure. . . '.   Each
amoeba in the spore mass [the fruiting body] now encases itself
in cellulose and becomes a spore.   The end result is a delicate
tapering shaft capped by a spherical mass of spores.   When the
spores are dispersed (by water, or contact with some passing
creature such as an insect or a worm) each can split open to
liberate a tiny new amoeba."

Thus the cycle of the community begins again, with the amoeba
population thriving and growing on their newly located bacterial
food, until this food supply is exhausted and the cycle of aggregation and culmination begins anew.   This microscopic, brainless
creature enacts in the course of its life history the parable of all
communities.   As they do in all communities, certain members
of this microcosmic society seemingly take upon themselves the
responsibilities of leadership, initiating the activities of the group.

Just how this happens, was demonstrated by a young Cambridge zoologist named Bryan M. Shaffer, who went to the
United States in 1956 and worked first at Bonner's laboratory in
Princeton and then in Raper's laboratory, which had by this time
been moved to the University of Wisconsin.   To prevent
aggregating cells from mounting one another in clumps, thus
obscuring what each cell was doing, Shaffer devised what he
called a "sandwich" technique, a complicated arrangement of oil
and water films floating between glass slides, which forced the
amoeba to aggregate in a single layer without clumping.   In
this way he discovered the existence of what he called "founder"
cells.   He wrote in his description that "a founder varied somewhat in appearance [from ordinary amoeba] at the time it became

active.   In many cases it was oval or almost completely circular
in outline and stationary. . . .   Because it was less expanded, it
frequently appeared smaller than most or even all of the cells that
responded to it, and also darker. . . .   It abruptly began to affect
its neighbors over a considerable area.   These elongated toward
it within a few minutes. . . .   The first to reach the founder began
to encircle it, either in one direction, or, becoming temporarily
Y-shaped, in both.   Whether it was able to surround it depended
partly on their relative sizes and partly on how soon further cells
arrived, for these competed for the founder's surface.   Such
intimate contact was established that a two or three-celled center
could sometimes be mistaken for a single giant cell.   Occasionally, perhaps five or ten minutes after starting to attract, a strong
founder became less circular and less refractive [dark] and stopped
attracting.   The responding cells became less elongated and
tended to produce pseudopodia from other parts of their surfaces;
and then perhaps five minutes later, the founder rounded up again
and the others moved toward it."

Shaffer wondered whether their role as founders was predetermined by their particular genetic heritage, or whether any
amoeba could become a founder, whether it was mere chance that
determined which amoeba would first feel the pinch of food shortage and become impelled to disseminate the signal for assembly.

The nature of this signal was chemical—a gas.   This had been
determined several years previously by Bonner, who had designed
a series of experiments which systematically excluded possible
nonchemical signals, such as electrical fields or other chemical
communication systems, like contact or molecular trails left by
amoeba in the course of their travels.   Though unable to isolate
the chemical molecule which comprised the gas, he named the
substance acrasin, a coinage from the technical class-name of the
animal which produced it.

There are no clear-cut conclusions to be reached as a result of
the experiments Shaffer subsequently conducted.   He seeded
a culture, permitted it to flourish for a period, and then deprived

it of food until a founder cell formed. "If this," he writes, "was immediately killed, and the culture at once returned to darkness, the residual cells did not re-aggregate. But if it was left in the light, a new founder eventually did appear."

However unhappily inconclusive this work may have been, the presumption remains from other evidence that varying different social roles may be assumed by many members of the population. As early as 1902 biologists were bringing slime molds back to their laboratories, and doing what almost any child would do—seeing what would happen if a migrating slug were cut into several sections. And what happened was this: Each section ceased its migration and promptly entered into the culminating stage, producing a fruiting body that was somewhat smaller than normal in size, but seemingly perfectly normal in all other respects. From its spores perfectly normal new populations would spring.

Among the slime molds the question of whether differences of behavior pre-exist in differing members of the population, or whether any behavior can be assumed by any member of the population, is still open. If the migrating slug is cut apart, new migration leaders will not appear. Migration as an activity ceases. But leaders that initiate other specialized social roles do appear. Stalk-building, a social role which would normally be assumed by the leadership cadre at the apical tip is now assumed by other amoeba quite distant from the tip. In the case of the slug which is cut into three sections, certain amoeba in the terminal section far removed from the apical tip will nonetheless commence filling themselves with water, encasing themselves in cellulose and beginning to migrate backwards through the mass of their fellows in order to form the stalk. Under normal circumstances these animals would not form the stalk, but would remain anonymous members of the sorocarp; each one would have transformed itself into a spore, not a bit of stalk.

It is also curious that in the slime molds the individual members of the community that form the leadership group are nonreproductive.   Only those animals that form themselves into spores contribute genetically to subsequent generations.   The amoeba at the apical tip seemingly perform a totally sterile role in the future of the community, by bearing the mass of their fellows aloft, raising the community away from contact with the earth, thus assuring the community more favorable possibilities for dispersion.   This is curious, for it has a relationship to human societies; it overleaps the evidence of most vertebrate communities where the dominant animals, particularly males, have greater sexual opportunities and generally pass on their genetic characteristics differentially—in a larger statistical degree—than low-ranking, non-dominant individuals.   Often in human societies many of the most valuable members are nonreproductive; members of celibate religious orders, ascetics, homosexuals, and so on.   Reproduction flourishes on the lower levels of societal competence, the marginal members often reproduce disproportionately large numbers of themselves.

Marvelously instructive as this parable of the slime molds will very likely be—in furthering our understanding of how cells aggregate to form the fetal stage of multicelled organisms, how our understanding of slime mold behavior can contribute to our understanding of human growth and wound-healing and many other related phenomena—it is somehow still surprising how little is known about the actual process itself.

In part the problems inherent in the study of slime mold sociology derive from the microscopic scale of observation; it is very difficult to distinguish individuals within the group, difficult to mark them as individuals and follow them in the course of their social activities.   No such scale difficulties involving technical problems impede our understanding of vertebrate sociology. There the problem is of a different order—one that may perhaps be considered to result from a barrier erected by human vanity. For it is vanity that identifies the self as unique and separate from all that is nonself.   The human mind, working through the eye, is a synthesizing instrument.   When we look, for example, upon a piece of cloth, we see a fabric and not an ordered collection of individual fibers. An earthenware pot exists in the mind of the human beholder as a plastic form in space and not a collection of particles of clay.

And so it was quite late that Western man came to understand that social orders exist in vertebrate communities other than his own. The order that exists in insect communities could not be denied. But humans observed schools of fish, flocks of birds, herds of antelope, and saw them as totalities in much the same disinterested way as one might view a cloud of insects around a streetlamp on a summer evening. No one apparently deemed it a fit concern of science to attempt to distinguish the individuals from within such a collective and study their interactions with other members of the aggregation. As far back as the Pleistocene period, human hunters doubtless perceived that very often such a collection of vertebrates as a herd of deer, or a school of fish, or a flock of birds might contain a leader, an individual who stimulated and directed the group's activities. Modern zoologists also realized this, but what they did not realize was that the leader arrived at his (or her) position by ascending a graduated ladder of social responsibility. Social recognition comes to animal leaders in much the same way as it does to human leaders, by the community's consent to the leader's competence. Within each stable animal society there exists a scale of interlocking social relationships and since the acquisition of leadership (or dominance) requires that these relationships be manipulated, animal leaders are equally as political as human leaders.

This fact, which is now considered obvious, was first noted by a Norwegian zoologist named Thorlief Schjelderup-Ebbe. It was in 1913 that he first announced to his parochial Norwegian community of zoologists that political hierarchies existed in barnyard chicken coops, and later (in 1922) he announced these findings in German to the world at large. He demonstrated to his fellow scientists by designing a series of hard, statistics-laden experiments, which were widely reproduced, the existence of a condition that had been well known to every farm boy since the domestication of a barnyard fowl. He demonstrated the obvious fact that every chicken and every rooster in a barnyard flock is an individual; and that the flock only becomes transformed into a society when the membership comes to recognize one another as individuals and acknowledges a graded hierarchy of social rank. Schjelderup-Ebbe writes of that peculiar vanity that blinds humans to the recognition of individuals. The human self only recognizes other, similar-appearing selves as being potentially unique, being individuals. Schjelderup-Ebbe begins one of his most important papers with the statement: "Every bird is a personality. . . . This may sound odd, but it is only because the individual and social psychology of birds has been regarded too superficially. No attempt has been made to know each individual bird in a given flock. So to know them, however, is the most important prerequisite for the full understanding of the general and comparative psychology of birds. The ability to distinguish each individual provides the key for the solution of a series of problems which we should other wise be unable to solve, and which are not only of ornithological interest but also of importance for the understanding of the general continuity that prevails in all life.

"It must be admitted that this ability to distinguish every single bird within a number of species presents various difficulties. These difficulties can, however, be overcome by careful observation. In some birds, for example the jackdaw (*Corvus monedula*), it is very difficult to distinguish the individuals of a group immediately; such discrimination, in fact, can be accomplished only by the most careful observation of minor points in which they may differ from one another, such as the size of the head, size of the eyes, etc. Repeated experience however, enables us to learn to recognize individuals, because in time—often quite unconsciously—we become more familiar with the characteristics of the species and thus more cognizant of slight variations. Individuals belonging to the races of men other than our own are distinguished with great difficulty, especially if we have not often had the opportunity of seeing members of this alien race."

From the point of view of his fellow scientists, Schjelderup-Ebbe had an unfortunate (most unscientific) capacity for empathy. This led him into the scientific pitfall of anthropomorphism. Despite this handicap his data was so well developed and his experiments so easily reproducible that his work convinced his scientific peers in spite of the purple prose in which he described it. For example, here is his description of the appearance of a hen after making an attack to enforce her dominant status: ". . . the face of the despot would radiate with the joy of satisfied pecking lust, and the fury could clearly be observed in its eyes. These expressions of fury we again find very clearly marked in birds in numerous other cases, for example the expression of intense spitefulness which several parrots, e.g., gray parrots (*Psittacus erithacus*) and the yellow-blue ara (*Sitace araruna*) display on their faces when they draw blood from human beings by vicious pecks. Gray parrots even attend this expression with peals of discordant laughter and other harsh vocalizations while other parrots follow it up by nodding of the head, violent flapping of the wings, and piercing screams."

Schjelderup-Ebbe saw the acquisition of social dominance in terms of despot and slave. He saw the formation of a social order to be as malevolent as any occurring in nature—in fact he saw all of nature to be malevolent. He wrote at one time: "Despotism is the basic idea of the world, indissolubly bound up with all life and existence. On it rests the meaning of the struggle for existence." In his day, there was nothing whatever exceptional in this point of view. He was a careful and stubborn observer of animals. He was also sufficiently sympathetic to them to have been one of the first professional zoologists to recognize them as individuals. With all this, however, he was by no means an original thinker and his work merely augmented that already existing pessimistic Malthusian view of the competitive universe.

Friederich Nietzche is reported to have said that "over the

whole of English Darwinism, there hovers something of the odor of humble people in need and in straits." This odor that Nietzche sniffed was the odor of overcrowding. Nineteenth-century England contained some of the world's worst slums; evidence testifying to the rightness of Malthus' view of nature abounded on every hand. There was little else to be seen except this terrible, competitive struggle for existence. Though Schjelderup-Ebbe was not an Englishman, his writings are steeped in the same unpleasant odor. His first insights into social hierarchies came from observing chickens, which in the course of their long evolutionary adaptation to domesticity, were forced to evolve stringent behavioral mechanisms to keep their overcrowded coops from degenerating into behavioral sinks.

While superficially it may seem paradoxical that a more balanced and benign view of evolution should have been promoted by a pair of aristocrats who emerged from highly authoritarian societies, they were nonetheless freed from one influence which cramped the thought of many of their contemporaries; they did not live their early lives in overcrowded cities. They lived on vast, sprawling, landed estates. Yet the paradox still remains, for Baron von Uexküll came from within a class of Prussian junkers that had systematically trained itself from womb to tomb in the suppression of all gentle, tender, and sympathetic sentiments. The other aristocrat, a Russian Prince, Pëtr Alekseyevich Kropotkin, emerged from that class in Russia that persisted as a class only through the exercise of the most total and tyrannical ruthlessness. Both these men had first-hand experience with social injustice, but from that side of the proscenium that would more easily permit the development of compassionate rather than vengeful feelings.

Because Kropotkin's views were so radically prophetic for his time, and because he was possessed by them, he suffered a life of poverty, imprisonment and exile: it is interesting to know some of the details of his life in order to comprehend how he came to be what he was.

Photographs taken of him early in life show a strikingly handsome young man with a melancholy brooding face. He sits in uniform at a table, his chin resting on his knuckles in a romantic pose; the whole arrangement is so stereotyped that the photograph could be that of any introspective, narcissistic aspiring actor. But the photographs taken of the old man are quite different. They show a robust figure, bald, with a fringe of white disarrayed hair over the ears, wearing a large, patriarchal white beard spreading like a fan over his chest. The most forceful impression one gets from the photograph is the transmitted feeling of peace and repose. The eyes observe the photographer through silly-small Ben Franklin spectacles, which sit crookedly across the bridge of a nose that seems to have been broken and badly mended. The love and compassion which stream from these eyes, even in a faded photograph, have a strangely compelling force. Kropotkin spent most of his mature life in London, in exile, and in grinding poverty. This was a striking contrast to the conditions of his birth and youth. His British biographer, George Woodcock, writes, "he was born into the highest rank of the Russian aristocracy. The Kropotkins had been princes of Smolensk . . . and were descended from the dynasty of Rurik which had governed Russia before the Romanovs." The elder Prince, Kropotkin's father, was described by a contemporary as a "coarse and stubborn landowner" who ruled his 1,200 serfs in a totally despotic manner. Kropotkin's older brother also refers to their father in a letter to Pëtr in London as being "nasty, revengeful, obstinate and mean." Though mean he may have been, he did not stint in the florid splendor of his household. For the family of six there was a servant retinue of over fifty persons including a tailor, a piano tuner, a confectioner, and a band of twelve musicians; all serfs.

At the age of thirteen, Kropotkin was awarded the highest honor available to a young nobleman, appointment to the Tsar's personal corps of pages. While at the school for pages, Kropotkin came under the influence of two men who were to shape his life: one of them was a German, a Colonel Winkler, who taught mathematics as an adjunct to his principal subject, the use of artillery. Winkler introduced the new system of importing guest lecturers for the pages from St. Petersburg University. The other was Klassovsky, a university classicist who marked the forming mind of Kropotkin with, as he himself said, "an immense influence, which only grew with the years."

It was at this stage that he became fascinated with geography and natural history. Recalling this period, he wrote: "The never ceasing life of the universe, which I conceived as *life* and evolution became for me an inexhaustible source of higher poetic thought, and gradually, the sense of man's oneness with nature, both animate and inanimate—the poetry of nature—became the philosophy of my life."   In 1862 Kropotkin ended his tour of duty with the corps of pages and was forced to choose a regiment to which he would be commissioned as a junior officer.   He writes: "My thoughts turned more and more toward Siberia. The Amur region had recently been annexed by Russia; I had read all about this Mississippi of the East, the mountains it pierces, the subtropical vegetation of its tributary the Usuri, and my thoughts went further—to the tropical regions which Humbolt had described and to the great generalizations of Ritter, which I had delighted to read.   Besides, I reasoned, there is in Siberia an immense field for the application of the great reforms which have been made or are coming; the workers must be few there, and I shall find a field for action to my tastes."

He asked to be commissioned to a regiment of Cossack cavalry stationed in the farthest reaches of Siberia near the Manchurian border.   Court circles were dismayed by this irregular choice of units, but by means of various machinations, Kropotkin managed to get his way.

On arriving in Siberia, he found his superior General Kukel, head of the General Staff, to be a congenial man.   Kukel had been a personal friend of Bakunin, the anarchist philosopher, who had just recently escaped from his Siberian prison, found his way

to China, and eventually Japan.   Kukel introduced Kropotkin
to Bakunin's wife; the three of them spent many hours together
and these influences also shaped Kropotkin's later life.   It is quite
likely that his later anarchist political activities stemmed from the
impressions made on him by those long evenings spent with
General Kukel and Madam Bakunin.

Kukel also encouraged Kropotkin's geographical explorations.
Tsarist officers were not popular in those remote areas, so Kropotkin made his initial voyages of exploration alone, unarmed, and
in disguise.   He designed an overcoat with extra large pockets
and trained himself to write surreptitiously, so that he could take
notes blindly, pencil, paper, and writing hand all concealed within
the capacious pocket.   He made five major journeys and a
number of smaller expeditions and covered over 50,000 miles of
unexplored territory mostly on horseback.   At one point he was
recalled from Siberia to make a personal report to his superiors
in St. Petersburg, and was forced to travel during the spring
thaw, the most treacherous time for crossing the tundra, 3,200
miles from Irkutsk to the railroad terminal at Novgorod.   He
managed to traverse this distance in twenty days, thus maintaining
an average pace of 160 miles a day, or ten miles an hour for
sixteen hours each day.   He was obviously a man of incredible
strength and fortitude.

In 1865 he was assigned his most important exploration party.
This time he was not to travel alone, but to take with him a zoologist, one Poliakov, and a topographer, Maskinski.   They were
to travel in an armed party of ten Cossacks and fifty horses.   He
writes of this experience as follows: "I recollect myself, the
impressions produced upon me by the animal world of Siberia
when I explored the Vitim regions in the company of so accomplished a zoologist as my friend Poliakov.   We were both under
the fresh impression of *The Origin of Species*, but we looked vainly
for the keen competition between animals of the same species
which the reading of Darwin's work had prepared us to expect. . . .
We saw plenty of adaptations for struggling, very often in common

against the adverse conditions of climate, or against various enemies, and Poliakov wrote many a good page upon the mutual dependency of carnivores, ruminants, and rodents in their geographical distribution; we witnessed numbers of facts of mutual support, especially during the migration of birds and ruminants; but even in the Amur and Usuri region where animal life swarms in abundance, facts of real competition and struggle between higher animals of the same species came very seldom under my notice, though I eagerly searched for them."

The principal purpose underlying Imperial sponsorship of the expedition was not primarily scientific but military; Kropotkin was expected to discover a direct means of communication between the Lena gold mines and Transbakalia. A large, elaborate expedition to seek a pass through the Sayan Highlands had been mounted in 1860 and had spent four fruitless years in the area. It had finally been recalled, and Kropotkin with his small party, it was hoped in St. Petersburg, would have better success. Kropotkin found a pass within the year. He was commended by the military authorities and he also prepared a paper, for presentation to the Royal Geographic Society, which was so well received that on the strength of this accomplishment, he was awarded the society's gold medal.

At this point Kropotkin resigned his commission and returned to St. Petersburg. He took a cheap and barren room in the students' quarter, where he began pondering inconsistencies, which he had noted in Humbolt's scheme of the historical geology of the Asian continent. He collected vast amounts of data from the library, from the journals of other travelers, until at last one day he experienced an intuition that enabled him to see the whole solution to the land formation of Eurasia clearly "as if it were illuminated with a flash of light." He wrote that "the main structural lines of Asia are not north and south, or west and east; they are from the southwest to the northeast—just as in the Rocky Mountains and the plateaus of America the lines are northwest to southeast; only secondary ridges shoot out northwest." The complete development of this theory occupied Kropotkin for several years, and in 1870 he wrote a paper imagining the existence of a land mass near Novaya Semla, which was generally believed to be an ice-covered sea not underlaid by land. Plans to mount an expedition to test this hypothesis aborted, and it remained to an Austrian polar expedition headed by Julius von Payer and Karl Weyprecht, which two years later advanced along the route suggested by Kropotkin and discovered the archipelago that Payer and Weyprecht named Franz Joseph Land, thus validating Kropotkin's theory.

With this accomplishment Kropotkin now enjoyed a world-wide reputation as a geographer; he was nominated for the presidency of the Physical Geography Section of the Russian Geographical Society, but unfortunately he was unable to assume the post because within hours after leaving the meeting at which he had presented his paper, he was arrested by the secret police. He was not yet thirty by the time he had made for himself an international reputation, but the acclaim meant little to him. The important thing for him was the moment of creative rapture about which he writes beautifully: "There are not many joys in human life equal to the joy of the sudden birth of a generalization illuminating the mind after a long long period of patient research. What has seemed for years so chaotic, so contradictory, and so problematical takes at once its proper position within a harmonious whole. Out of the wild confusion of facts, and from behind the fog of guesses—contradicted almost as soon as they are born—a stately picture makes its appearance, like an Alpine chain suddenly emerging from the mists which concealed it the moment before, glittering under the rays of the sun in all its simplicity and variety, in all its mightiness and beauty. . . .

"He who has once in his life experienced this joy of scientific creation will never forget it; he will long to renew it; and he cannot but feel with pain, that this sort of happiness is the lot of so few of us, while so many could also live through it—on a small or on a grand scale—if scientific methods and leisure were not limited to a handful of men."

While he was engaged in preparing this paper outlining his general theory of the historical development of the land mass of Eurasia, he had also become involved with a group of intellectuals who gave covert lectures to workers' groups on political and also on nonpolitical subjects. Kropotkin, adopting the pseudonym Borodin, and dressed in peasant disguise, addressed several of these meetings. Stepniak, another lecturer in the same group, recalls Kropotkin's appearances: "These lectures to which a depth of thought united a clearness and a simplicity that rendered them intelligible to the most uncultivated minds, excited the deepest interest among the working men of the Alexander Nevsky district. They talked about them to their fellow workmen and the news spread quickly through all the workshops of the neighborhood and naturally reached the police who determined at all hazards to find out the famous Borodin."

A week before Kropotkin was due to deliver his paper to the Geographical Society, he became aware that he was under surveillance by the secret police. He was determined to brazen it out if he could, in the hope that he might be able to read his paper in person and then flee the country following the meeting. He was watched at the meeting but no move was made to apprehend him until after he had returned to his room and made a clumsy and surprisingly inept attempt to escape unnoticed down the backstairs.

He was arrested, interrogated for several days (during which he revealed nothing), and then taken to the fortress of Peter and Paul where he was formally introduced to the governor of the prison, then stripped, given a dressing gown and a pair of slippers, and installed in a dank cellar dungeon where he remained in totally solitary confinement for two years. From the beginning of his imprisonment he was allowed access to the inadequate prison library and to tobacco. He immediately set up a routine for himself to preserve his sanity. Every moment of the day was used: he read, and meditated, and performed various exercises, walking a minimum of five miles a day back and forth across his cell, and inventing various drill routines using a heavy stool in his cell in lieu of barbells.   In this way he was able to maintain his sanity and preserve his health for quite a long time—for two years, in fact—until at last he succumbed to scurvy.   He was removed to another prison where he was put in a cell with other men, but he received no treatment and his health continued to decline to such a point that his warders feared for his life.   He was then removed to the Military Hospital at St. Petersburg from which place he resolved to escape.

More than twenty people were finally involved in the daring escape plan.   One lookout was installed in a room near the prison hospital.   His violin was audible to Kropotkin walking in the hospital yard and when the coast was clear he was instructed to play a mazurka.   Another lookout (who proved on the final day to be superfluous) was detailed to sit upon a stone and eat cherries. He would stop eating when a threat appeared.   Kropotkin could presumably see the working of his masseter muscles from within the courtyard.   In the hospital (partly through the mediation of the Russian learned societies with the Tsar) Kropotkin was allowed visitors and the plan was able to evolve.   Final details were written down and smuggled to him in the false back of a pocket watch.

The initial phase required Kropotkin to sprint 150 yards across the yard to the main gate.   On the afternoon of the pre-arranged day, Kropotkin came into the yard at four o'clock.   He himself writes what followed: "Immediately the violinist—a good one, I must say—began a wildly exciting mazurka from Konsky as if to say 'Straight on now—this is your time!'   I moved slowly to the nearer end of the footpath trembling at the thought that the mazurka might stop before I reached it.

"When I was there, I turned round.   The sentry had stopped five or six paces behind me: he was looking the other way. 'Now or never!'   I remember that thought flashing through my head.   I flung off my green flannel dressing gown and began to run.

"For many days in succession, I had practiced how to get rid of that immeasurably long and cumbersome garment.   It was so long that I carried the lower part on my left arm as ladies carry the trains of their riding habits.   Do what I might, it would not come off in one movement.   I cut the seams under the armpits, but that did not help.   Then I decided to throw it off in two movements; one, casting the end from my arm, the other dropping the gown on the floor.   I practiced patiently in my room until I could do it as neatly as soldiers handle their rifles.   'One, two' and it was on the ground.

"I did not trust much to my vigor and began to run rather slowly to economize my strength.   But no sooner had I taken a few steps than the peasants who were piling the wood at the other end shouted, 'He runs!   Stop him!   Catch him!' and they hastened to intercept me at the gate.   Then I flew for my life.   I thought of nothing but running—not even of the pit which the carts had dug out at the gate.   Run! run! full speed!

"The sentry, I was told later by friends who had witnessed the scene from the grey house, ran after me, followed by three soldiers who had been sitting on the doorsteps.   The sentry was so near to me that he felt sure of catching me.   Several times he flung his rifle forward trying to give me a blow in the back with the bayonet.   One moment my friends in the window thought he had me.   He was so convinced that he could stop me in this way that he did not fire.   But I kept my distance and he had to give up at the gate.

"Safe out of the gate, I perceived to my terror that the carriage was occupied by a civilian who wore a military cap.   He sat without turning his head to me.   'Sold' was my first thought.   The comrades had written in their last letter, 'Once in the street don't give yourself up.   There will be friends to defend you in case of need,' and I did not want to jump into the carriage if it was occupied by an enemy.   However, as I got nearer to the carriage,

I noticed that the man in it had sandy whiskers[1] which seemed to
be those of a warm friend of mine. He did not belong to our
circle, but we were personal friends, and on more than one
occasion, I had learned to know his admirable, daring courage,
and how his strength suddenly became herculean when there was
danger at hand. 'Why should he be there? Is it possible?' I
reflected, and was going to shout out his name when I caught
myself in good time and instead, clapped my hands while still
running to attract his attention. He turned his face to me—and I
knew who it was.

"'Jump in quick, quick,' he shouted in a terrible voice, calling
me and the coachman all sorts of names, a revolver in his hand
and ready to shoot. 'Gallop! Gallop! I will kill you!' he
shouted to the coachman. The horse—beautiful racing trotter
which had been bought on purpose—started off at full gallop.
Scores of voices yelling 'hold them! Get them!' resounded from
behind us, my friend meanwhile helping me into an elegant
overcoat and an opera hat. But the real danger was not so much
in the pursuers as in a soldier who was posted at the gate of the
hospital about opposite to the spot where the carriage had to wait.
He could have prevented my jumping into the carriage, or could
have stopped the horse by simply rushing a few steps forward. A
friend was consequently commissioned to divert this soldier by
talking. He did this most successfully. The soldier, having
been employed at one time in the laboratory of the hospital, my
friend gave a scientific turn to their chat, speaking about the
microscope and the wonderful things one sees through it.
Referring to a certain parasite of the human body, he asked, 'Did
you ever see what a formidable tail it has?'

[1] A rumor circulated, for some time in court circles following Kropotkin's
escape. The rumor claimed that the man with the sandy whiskers in the military
cap was none other than Grand Duke Nicholas, brother of the Tsar. The Grand
Duke did have sandy whiskers and was known to have visited Kropotkin in the
hospital. The uncharacteristic coyness with which Kropotkin writes ". . . and
I knew who it was" possibly lends credence to this story. This rumor also makes
it more credible that this over-elaborate plot functioned with such efficiency.

"'What man, a tail?'

"'Yes it has.   Under the microscope it is as big as that.'

"'Don't tell me any of your tales,' retorted the soldier, 'I know better.   It was the first thing I looked at under the microscope.'

"This animated discussion took place just as I ran past them and sprang into the carriage.   It sounds like fable, but it is fact."

By foresightedly planning for other conspirators to hire all the horsecabs in the area—within the radius of a mile of the hospital—the plotters managed to forestall any effective direct pursuit of Kropotkin's carriage.   He made a clean escape.

He shaved his beard, was provided with an officer's uniform on the assumption, which later proved quite correct, that in despotic Russia, customs agents, border guards, etc., would be fearful of incurring the displeasure of an officer by delaying him with an overly scrupulous examination of his papers.   Kropotkin crossed the Russian border into Finland, from Finland he took ship to Sweden and then to Norway, where he booked sea passage to Hull, England, where he landed at last, wearing the name of Alexis Levashov.   Suspecting that Russian secret agents would expect him to go immediately to London, and fearing assassination at their hands, he rented a room in Edinburgh.

After some time, hearing from friends in Russia that the search had moved to Lausanne, Switzerland, and (oddly enough) to Philadelphia, Pennsylvania, Kropotkin felt sufficiently secure to move to London, where he looked up James Scott Keltie, the assistant editor of the British journal, *Nature*.   He began working for Keltie (the two became good friends eventually) translating items from the foreign journals into English.   Finally one day he was forced to reveal his true identity, when Keltie gave him his own book on the glacial history of Eurasia to review.   Keltie later wrote about the incident with proper British emotional economy: "I told him we had no one in a position to review the book and he might write an article stating briefly its main features and conclusions, which I am glad to say, he did."

It was while working for Keltie that Kropotkin became

increasingly interested in biology, particularly in animal social behavior. In 1877 the French zoologist Alfred Espinas published his book *Les Societes Animales*, which set many ideas to churning in Kropotkin's head. In 1880 Kropotkin came into possession of a copy of an address delivered by the Russian zoologist, Karl Fedorovich Kessler, Dean of St. Petersburg University, on the subject of animal co-operation as a factor in evolution.

It was eight years later, however, before Kropotkin was finally stimulated into formulating and articulating his own views. The trigger was the publication of a paper by Thomas Huxley entitled "The Struggle for Existence and Its Bearing Upon Man." In this paper Huxley extrapolated from the observations of the neo-Darwinians who saw all about them evidence of prey-predator violence among differing species of animals, and concluded that the same conditions must have prevailed among primitive humans. "Life was a continuous free fight," wrote Huxley with great and confident certitude, "and beyond the limited and temporary relaxations of the family, the Hobbesian war of each against all was the normal state of existence."

Kropotkin was appalled. Voicing his disagreement to his friends, he was encouraged by the British zoologist Henry W. Bates, and by John Knowles, editor of *The Nineteenth Century*, to prepare a rebuttal. A series by Kropotkin titled "Mutual Aid Among Animals, a Factor in Evolution" began appearing in serial form in September 1890 in Knowles' magazine. In his introduction to the book, which finally appeared as a collection of these essays, Kropotkin writes: "Two aspects of animal life impressed me most during the journeys which I made in my youth in Eastern Siberia and Northern Manchuria. One of them was the extreme severity of the struggle for existence, which most species of animals have to carry on against an inclement Nature; the enormous destruction of life, which periodically results from natural agencies; and the consequent paucity of life over the vast territory which fell under my observation. And the other was, that even in those few spots where animal life teemed in abundance,

I failed to find—although I was eagerly looking for it—that bitter struggle for the means of existence *among animals belonging to the same species*, which was considered by most Darwinists (though not always by Darwin himself) as the dominant characteristic of the struggle for life, and the main factor of evolution.

"The terrible snow storms, which sweep over the northern portion of Eurasia in the later part of the winter, and the glazed frost that often follows them; the frosts and snow storms, which return every year in the second half of May when the trees are already in full blossom and insect life swarms everywhere; the early frosts and occasionally, the heavy snowfalls in July and August, which suddenly destroy myriads of insects as well as the second broods of birds in the prairies; the torrential rains, due to the monsoons . . . and finally the heavy snowfalls in October, which eventually render a territory as large as France and Germany absolutely impracticable for ruminants and destroy them by the thousands—these were the conditions under which I saw animal life struggling in Northern Asia. This made me realize at an early date, the overwhelming importance of what Darwin described as the 'natural checks to overpopulation' in comparison to the struggle between individuals of the same species for the means of subsistence, which may go on here and there to some limited extent, but never attains the importance of the former. Paucity of life—underpopulation—not overpopulation being the distinctive feature of that immense part of the globe, which we name Northern Asia. I conceived since then serious doubts—which subsequent study has only confirmed—as to the reality of that fearful competition for food and life within each species, which was an article of faith with most Darwinists, and consequently as to the dominant part, which this sort of competition was supposed to play in the evolution of new species."

It is true not only of animal populations, but of human populations as well, that when hostile circumstances press upon the community, they seem only to strengthen the communal bonds of co-operation. A large part of that sentimental nostalgia that attaches to the American frontier stems from the sense of communal solidarity that prevails in any frontier settlement. One can even imagine that human societies began to evolve in response to just those frightful kinds of climactic conditions described by Kropotkin, which must have prevailed during the succession of glacial periods of the Pleistocene, during which mankind and its early social institutions were formed.

It is unfortunate that in the recent temperate-zone history of mankind, the mold of crisis, which casts societies into cohesive forms, was not so much provided by the natural environment, as it was by the artifice of human war. Several times in this century we have seen nationalist politicians flay their peoples into communal frenzies by creating war threats for the express purpose of welding the mass of individuals into a cohesive unit.

Strong remnants of neo-Darwinian logic still persist in the doctrines espoused by various military philosophers. Terrorism is still believed (according to political pragmatists) to be an effective device for demoralizing an enemy. Though this experimental situation has been tried again and again during this century, it has not yet succeeded in its hoped-for effects. None of the major population centers, bombed so mercilessly by both sides (with conventional explosives) in World War II, collapsed into demoralization as a result of that bombing. One might call it the "Job Effect"—the more a community is afflicted by suffering, the more it seems sustained by this affliction, the more capable it seems to be of sustaining still more, right up to the point of total physical destruction. It seems that in cases of disaster, either man-made or natural, that the communal nature of communities is reinforced, given a new and viable vitality.

Kropotkin points out that Darwin was very well aware of the evolutionary importance of mutual aid. He quotes Darwin as having written that under adverse conditions, "those communities which included the greatest number of the most sympathetic members would flourish best and produce the greatest number of offspring. But," Kropotkin continues, "it happened with

Darwin's theory as it always happens with theories having any bearing upon human relations. Instead of widening it according to his own limits, his followers narrowed it still more. . . . They came to conceive of the animal world as a world of perpetual struggle among half-starved individuals thirsting for one another's blood. They made modern literature resound with the war cry of 'Woe to the vanquished.' "

Kropotkin, despite the power of his personal pacific convictions, was perhaps more able to view nature dispassionately than any of his surviving contemporaries. He took Huxley to task for the one-sidedness of his Hobbesian views; he also took Rousseau to task for what Kropotkin considered saccharin untruths contained in Rousseau's literary attitude toward nature. "It may be remarked at once," Kropotkin writes, "that Huxley's view of Nature had as little claim to be taken as scientific deduction as the opposite views of Rousseau, who saw in nature but love, peace, and harmony destroyed by the accession of man. In fact, the first walk in the forest, the first observations upon any animal society, or even the perusal of any serious work dealing with animal life . . . cannot but set the naturalist thinking about the part taken by social life in the life of animals, and prevent him from seeing in Nature nothing but a field of slaughter, just as this would prevent him from seeing in Nature nothing but harmony and peace. Rousseau has committed the error of excluding the beak and claw fight from his thoughts; and Huxley has committed the opposite error; but neither Rousseau's optimism, nor Huxley's pessimism can be accepted as an impartial interpretation of Nature. . . .

"As soon as we study animals—not in laboratories or museums, but in the forest and the prairie, in the steppes and mountains— we at once perceive that though there is an immense amount of warfare and extermination going on amidst various species and especially amidst various classes of animals, there is, at the same time, as much, or perhaps even more, of mutual support, mutual aid, and mutual defense amidst animals belonging to the same species, or at least to the same society.   Sociability is as much a law of nature as mutual struggle.   Of course it would be extremely difficult to estimate, however roughly, the relative numerical importance of both these series of facts."

The difficulty enumerated by Kropotkin in that last sentence soon proved, of course, to be the principal difficulty impeding widespread acceptance of his views at that time.   Though they appealed to all men of good will and intelligence, his writings lacked that solid statistical support which steadies the structure of any scientific theory.   Schjelderup-Ebbe could count the number of attacks made by one hen upon another within any given span of time, count the number of pecks delivered in each attack, and when it was all over, measure the length, depth and gravity of the wounds. An attack is an attack—but what is an act of "mutual aid"?

It was an American zoologist, Warder Clyde Allee, who finally came to Kropotkin's defense.   Though he had read Kropotkin while still in high school, and though he admired the Russian, it was not Kropotkin whom he tried to satisfy, it was himself. Allee was not a philosopher.   He was a good, kind, gentle, and very courageous man.   He was struck down in mid-career by a series of spinal operations necessitated by a birth defect, which left him eventually paraplegic.   He lived in a wheelchair and was taken out into the field riding piggyback on one of his sturdier students.   Yet if it did anything, his suffering seemed to accelerate his activity.   Though crippled, he accepted the chairmanship of the zoology department at the University of Chicago, handled all the burdensome administration, supervised the work of his graduate students, maintained a full teaching load, and in addition to all this, performed some of the most important experiments of his day, affirming, for all the world to see in the most indisputable terms, the truth of Kropotkin's insights.   He wrote over 300 technical papers, and eight books including the role of principal collaborator of one text that is still the classic, an 834-page, double-column tome titled *Principles of Animal Ecology*, which has gone through six printings to date.

He was a committed Quaker, having been born of a Quaker family; he married a Quaker woman in a Quaker meeting house and all his life was active in the Society of Friends.

Unlike many philosophers whose dreams, by the very scope and grandeur of their dimensions, obliterate detail, Allee was entranced by detail; he dealt with it as lovingly as a watchmaker.

It is unlikely that Allee ever came directly under the influence of Charles Whitman. He arrived at the University of Chicago as a nineteen-year-old graduate student, two years before Whitman's death. But he spent his summers at Woods Hole Biological Station, where he became attached to Frank Lillie (the marine worm watcher), who had become Whitman's successor as director of the Station. Whitman casts a long shadow over the course of modern zoological thought. Wherever one looks in the literature, one finds the most original and stimulating work being done by men who studied directly under Whitman, or by men who were affected by him at second hand. It is strange that Whitman himself left no taste of his character in the words he wrote. His papers are bone-dry and boring, making it very difficult to see him possessing the charismatic personality he must have had. He instilled in all the men who were affected by him a devoted attention to the minutiae of experimental detail and to the devising of dramatic demonstrations of basic principles.

If Kropotkin had been right, if "mutual aid" were in truth an evolutionary force, it must be, like all evolutionary forces, blind. It must simply be a force, like the wind, which promotes drift in a certain direction. It could not be a function of individual intelligence. Allee began his work by attempting to prove the existence and direction of such a force.

One of his early and most convincing experiments demonstrating the existence of just such a blind, thrusting force favoring aggregation over isolation was performed at Woods Hole Marine Biological Station. He used for his test animal the common sea urchin (that ball of prickly spines, the bane of coastal bathers), *Arbacia*. He writes that long before his arrival at Woods Hole,

Arbacia had "been much used in studies of various aspects of development. . . . There are several reasons for its popularity. These urchins are abundant in nearby waters and are readily mopped up by the tubful: they can be kept in good condition for some days in the float cages, and eggs and sperm are readily procured as needed.   Also the breeding season of *Arbacia* extends through July and August, which are favored months for research at the seaside.

"For years biologists at Woods Hole have studied the embryology and physiology of developing sea urchin eggs.   They have built up a painstaking, almost a ritualistic technique for handling glassware, towels, and instruments.   The procedures require as rigid a cleanliness as does a surgical operation.   Consequently, it was not surprising when I first took up the study of *Arbacia* to have one of my frankest friends among the long-time workers on their development voice what was apparently a common feeling. He asked pointedly whether I thought I could come into that well worked field and, without long training, find something he and his associates had overlooked."   Allee was not looking for something they had overlooked; whether he chose to admit it or not, he was searching for tangible evidence of an abstraction—he was searching to support his Quaker faith.

"The shed eggs of *Arbacia*," he continued, "are about the size of pin points and are just visible to the naked eye.   The spermatozoa are tiny things; the individual sperm are invisible without a microscope although readily seen when massed in large numbers. When a few drops of dilute sperm suspension are added to well washed eggs, one spermatozoan unites with one egg.

"After some fifty minutes at usual temperatures, the egg divides into two cells.   We call this the first cleavage.   Thirty or forty minutes later, a second cleavage takes place, and thereafter, cleavages occur rapidly.   Within a day, if all goes well, such an egg will have developed into a freely swimming larva."

By measuring cleavage time in sea urchin eggs, Allee was able to prove conclusively that the very process of growth itself—the process of cell division, one of the most basic biological processes anyone could possibly measure—was accelerated by aggregation. It proceeded (at least as measured by a linear time scale) in groups more readily than it did in isolation. He himself tells how he demonstrated this: "With appropriate experimental precautions, some 1,800 eggs were introduced into a tiny drop of sea water. Nearby on the same slide, forty eggs were placed in a similar drop, and the two connected by a narrow strait." In the illustration accompanying his paper, the two drops with their connecting strait look like a circus strongman's dumbbells. "A few eggs from the larger mass spilled over into this strait. The whole slide was placed in a moist chamber to avoid drying and was examined from time to time. In a trifle over fifty-five minutes, half the eggs in the densest drop had passed their first cleavage. A half minute later, 50 per cent of those in the strait were cleaved, and twenty seconds later, half of the more isolated ones had divided. The time to 50 per cent cleavage ranged between eighty-four minutes for the crowded eggs and over eighty-six and a half minutes for the isolated eggs."

He writes that as he repeated the experiment in differing ways with differing concentrations of eggs in each drop, he could finally detect a difference in cleavage time when only "sixty-five or more eggs were present in the more crowded drop and twenty-four or fewer eggs made up the accompanying sparse population." Since this original experiment was performed, the phenomenon has been thoroughly investigated, and the underlying mechanics are understood. Cleaving eggs release a protein fraction into the surrounding medium which serves as a stimulus to further cleavage; aggregations of eggs therefore are stimulated to cleave more rapidly by the reinforcement of chemical accelerators which they themselves produce, creating a kind of snowball effect. The precise mechanism in this instance is not quite so important as the demonstration of a principle, for Allee went on to prove the existence of this principle again and again, using different test animals and discovering different mechanics each

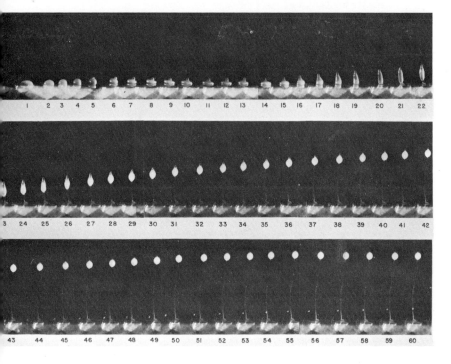

Dr. John T. Bonner's stop-motion series of the culmination stage of the society. The numbers are for reference only, they do not represent time spans (*Courtesy of Dr. John T. Bonner*)

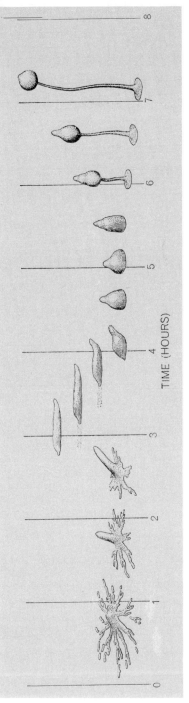

TIME (HOURS)

From aggregation to culmination, this drawing represents the life cycle of the society. Times indicated are only approximate—the migration stage, shown here as lasting merely one hour (from the third to the fourth hour) may be lengthened to cover a span of several days under certain conditions (*from How Slime Moulds Communicate by John Tyler Bonner. Copyright © 1963 by Scientific American, Inc. All rights reserved.*)

time. Though the mechanical system varied, it persistently promoted that inexorable "drift" that favored either the growth or the survival of animals in aggregations over animals existing in solitude.

For example, Allee performed a famous demonstration using a more biologically complex animal, a vertebrate form, a fish—the ordinary goldfish in fact. "Colloidal silver," he writes, "that is, the finely divided and dispersed suspension of metallic silver is highly toxic to living things, even the hardy goldfish.

"In the experiment in our laboratory, we exposed a set of ten goldfish in one liter of diluted silver, and at the same time placed each of ten similar goldfish in a whole liter of the same strength of the same suspension. This was repeated until we had killed seven lots of ten goldfish and their seventy accompanying, but isolated fellows. Without exception, the mean survival time of the group exceeded that of the accompanying isolated fishes." The goldfish which lived in groups survived in their poisonous medium for an average duration of 507 minutes, while the solitary goldfish survived an average time of only 182 minutes. This was a striking and unmistakable demonstration.

Upon searching for the underlying mechanics, Allee discovered that much of the silver in the beakers containing groups of ten fish had precipitated out of solution and lay as dust on the floor of the container. In the beakers containing solitary fish, almost all the silver was still suspended. He discovered that the fish secreted a slime from between their scales which caused the silver particles to drop out of the solution, and that the greater the number of fish present in any given amount of water, the greater the amount of slime secreted, and the more silver precipitated harmlessly down to the bottom of the tank before doing harm.

In a series of experiments designed to follow the *Arbacia* up the ladder of evolutionary complexity, Allee discovered that goldfish grow faster in groups than they do in isolation. Through either respiration or secretion of body substances, they "condition" the water in some still-mysterious fashion that promotes growth.

Goldfish are also sloppy eaters, regurgitating food in belches, which food is ingested by others in the group before it settles to the floor of the aquarium, so that even though in his experiments the food was carefully measured (Daphnia crustaceans were the food; they were sieved so that they were of similar size, and counted one by one so each portion was exact), group-living goldfish made better use of it. There is also the phenomenon of group stimulation: animals eating in groups seem to have their appetite stimulated by watching others feed.

Noting a report in the scientific press that mice seem to grow faster in groups than they do in isolation, Allee repeated the experiment in his laboratory hoping to find the operative mechanism. He did; in this instance it was thermostatic control. By their habit of huddling together, mice in groups share their body heat. Individuals isolated and prohibited from huddling expend energy in maintaining body heat, which in the case of mice reared in groups goes into growth.

At the same time that he was either conducting himself, or supervising experiments to show the benign physiological effects of group life over isolation, he also conducted experiments to show how group living facilitated learning and memory. The evidence produced by his experiments seems to be confusing, perhaps because of the emphasis on Watsonian "conditioning" techniques, which were popular at the time. Neither fish, nor insect, nor bird learned a task any faster as a member of a group than it would as an individual—but when a conditioned individual was introduced into an untrained group, the entire group learned the task far more swiftly and retained the memory of it longer than did control groups. Instead of trial-and-error discovery, untrained animals followed the leader in performing the task leading to reward.

Though Kropotkin is now generally disregarded as a serious zoologist theorist, since his surmises lacked the hard substance of statistical evidence once this evidence had been produced by Allee and others, evidence testifying to the adaptive usefulness of sociality, of group living as opposed to isolated living, the *principle* is still no clearer now than it was when Kropotkin first enunciated its existence.

It is much like the *principle* of animal mobility: we know that most animals are mobile at least at some time during their lives (though many molluscs and hydroids spend their adult lives lacking mobility). But though we know the *principle* of mobility is somehow overwhelmingly important, the mechanisms vary from the ciliary propulsion of the paramecium, to the jet propulsion of the squid, to the crawling of snakes, the flight of birds, and the walk of men. Just as no one has been able to find a common mechanism underlying the mobility *principle* of animal life (as opposed to the immobility of botanical life), Allee and his successors have likewise been unable to find a common mechanism underlying the sociality principle. All we know is what Kropotkin told us; that it is a need, and an overwhelmingly important one.

[8]

Aggression

It has long seemed a paradoxical inconsistency to poets and psychiatrists that God placed the site of love's pleasure so adjacent in the geography of the body (and in the geography of the infantile mind) to the site of excretion. This is true of almost all vertebrates, be they fish, reptiles or birds. And for mammalian males, the confusion is extended by the dual use of the same organ which on one occasion sows the fluid seeds of life, and on another serves to empty the body of dead substances.

Just as this paradoxical situation developed in the physiology and psychology of carnal love, another equally paradoxical equation of opposites developed as the physiology of *social* love evolved. Physiologically, insofar as the endocrine activities (which for many lower animals dominate behavior) are concerned, love runs concurrent with aggression. It has been somehow arranged for all vertebrates, that the formation of stable social units rests upon the exercising of what seems to us humans to be a totally separate and opposite drive—which we call aggression— a set of hostile feelings expressed in action which develops between conspecific members of the same group. The development of these feelings (and their resolution by what Konrad Lorenz has come to call *aggression suppressors*) are absolutely essential to the formation of stable social units. In fact, Lorenz and others have demonstrated that the degree of stability of any given unit can be directly correlated with the degree of aggression displayed during its formation. Lorenz even goes so far as to say in his book *On Aggression*, that "we do not know of a single animal which is capable of personal friendship and which lacks aggression. This combination is particularly impressive in animals that are aggressive only during the reproductive season, and which otherwise lack aggression and form anonymous flocks. When such creatures form any personal ties, they are dissolved with the loss of aggression. For this reason in storks, chaffinches, cichlids and others, the mates do not remain together when the big anonymous flocks assemble for migration." In the wild state, since many animal societies are stable the year round (such as baboon troops) this aggression is not necessarily acted out by the adult members of the society. It is literally "played out" by the young animals in scrapping quarrels and mock fights long before they reach physiological and social maturity.

This paradox is not nearly so apparent among invertebrates. Insect communities, for example, exhibit no intramural aggression in which one member pits itself against another member of the same community. And in those wilder reaches of animal social organization where interdependence reaches its most elaborately contrived evolutionary forms, among the hydroids and particularly in the subgroup known as the siphonophores, there is none at all. In that siphonophore colony we call the Portuguese man-o'-war, the transparent blue bladder that floats on the surface of the sea is one animal; each stinging tentacle that hangs from this float is another, each leech-like feeding polyp still another, and the community is further enlarged by separate male and female reproductive polyps. None of these animals can exist apart from the community and the community exists through the co-ordinated and collective labors of each class of members. The stinging polyps entrap and paralyze the prey animals, which serve the colony for food. The feeding polyps suck them dry of nourishing fluids, and this food, which they ingest, is shared throughout the community by an intricate system of internal food-sharing. While food-sharing rituals among the social insects has its basic adaptive purpose of communication, the

primary purpose of food-sharing among the members of a siphonophore colony is nutritive.   In the life style of this community,
Kropotkin's doctrine of mutual aid reaches its apogee of development.   There is nothing for any member of this community but
the total and utter dependence upon the community for mutual
aid.

The Portuguese man-o'-war, however, is not a society; it is a
colonial organism—at least this is the name given by zoologists
to this kind of co-operative living.   Those aggregations with
which Allee worked when he did his growth experiments with
Arbacia eggs, or with a haphazard collection of ten goldfish
arbitrarily enclosed for an arbitrary period within a beaker of
water; those too were not societies.   Neither is a swarm of
mayflies gathered around a streetlamp on a summer evening.   At
least we do not call these aggregations of animals societies.   A
society is defined as existing through the result of those interactions between the members comprising it, actions they perform
as individuals.   These interactions must be crucial to the maintenance and coherence of the society quite apart from any confining (the beaker) or attracting (the streetlamp) element in the
general environment.   Unlike invertebrate societies, such as
those of insects, vertebrate societies are held together by a thread
of tension, which binds every member to every other member.

This tension consists of that same strange alternation of drives
that promotes speciation—the drive to become part of that society,
part of a communion, and the drive toward individuation, the
drive to be particular and unique and to be so recognized.   Social
aggression is the means by which an animal expresses this latter
drive; the animal becomes aggressive in defense of its personal
space—its property, the extension in space of its person.   The
territory of fish and birds is markedly three-dimensional, a kind
of aura that the animal must perceive as being infused with itself.
Within this territory it is invariably dominant over other animals
that may invade it.

Ethologists, studying animals in the wild, have discovered

that there are several different types of aggression, social aggression, aggression in response to fear when escape is impossible, and that necessary kind of aggressive action that may be involved in their predatory food gathering. Animals manifest quite different behaviors in each of these instances; and social aggression also may differ markedly from instance to instance. Aggressive behavior often varies depending on whether it is directed against another animal as a result of male-male interactions, male-female, or adult-offspring. Because of some strange, self-imposed blindness, American behaviorist psychology failed to make these distinctions; aggression was considered to be a stimulus-response chain, and only the situational context that precipitated an aggressive act was studied. If a rat tried to bite its handler, or another rat, or even a cat, it was all the same to the behaviorist. It was merely an aggressive response to a hypothetical stress stimulus. But as ethologists began watching animals in the wild, they began seeing that aggression rarely produced an actual physical attack; they discovered what they called "intention movements" those preliminary gestures that aid an animal both physically and emotionally to switch from one state to another. These intention movements have come to serve as a communication system; they are often referred to in the literature as signals.

For example, many mammals use their long canine teeth not only for making puncture wounds, for catching a fold of flesh and biting through it, they also use these teeth as slashing, ripping weapons. In order to prepare themselves for an attack, they bare the teeth, unsheath them as it were, and like the hand of a swordsman half pulling the sword from the scabbard, this movement is often enough to alter the social situation so that no actual violence need occur. Only a tiny, almost imperceptible movement of the lip is all that is needed in some situations; the swordsman need not actually unsheath his weapon, often merely a significant movement of his hand in the direction of the swordhilt may be enough to resolve, or at least stabilize the situation.

In some species (as disparate as the cormorant and the lizard, which have both evolved similar structures, an inflatable, brightly colored air-sac under the throat) the aggression signal may be far removed from any functional intention movement. In other species, the intention movement has become so ritualized that it is no longer immediately apparent as such to a human observer. The seagull's act of rising up—stretching its neck tall, while at the same time tucking in its beak toward its chest, also half opening its wings, and perhaps flapping them in this half-opened position —is an intention movement. The seagull makes a physical attack upon an opponent pecking down at him while clubbing him with the heavy carpal joints of his wings. Any gull performing this intention movement is announcing a state of being or mood which alerts other gulls in the colony who read the signal precisely for what it is, a preparation for battle. The baring of the teeth by the carnivore still contains an unmistakable meaning for the human observer who may then choose to act accordingly, but the intention movement of the seagull is not quite so immediately comprehensible.

While ethology is not a new discipline, the name and need for it having been proposed by St. Hilaire in 1859, one of the most convincing and articulate apologists for this system has been a Viennese zoologist named Konrad Zacharias Lorenz, the son of a professor of orthopedic surgery at the University of Vienna. Adolph Lorenz loved the country and nature, and maintained a country house at Altenburg, where the family spent as much time as it could afford. Konrad and his older brother Albert explored the countryside. Konrad describes it in the following way: ". . . the virgin wildness of this stretch of country is something rarely found in the very heart of old Europe. . . . Protected against civilization and agriculture by the yearly inundations of Mother Danube, dense willow forests, impenetrable scrub, reed-grown marshes and drowsy backwaters stretch over many square miles; an island of utter wildness in the middle of lower Austria; an oasis of virgin nature, in which red and roe deer, herons and cormorants have survived the vicissitudes even of the last terrible war. . . . There is a strange contrast between the character of the landscape and its geographical situation and, to the naturalists' eye, this contrast is emphasized by the presence of a number of American plants and animals which have been introduced. The American goldenrod (*Solidago virgoarurea*) dominates the landscape above water as does *Elodea canadensis* below the surface: the American sun perch (*Eupomotis gibbosus*) and catfish (*Amiurus nebulosus*) are common in some backwaters; and something heavy and ponderous in the figure of our stags betrays, to the initiated, that Francis I in the heyday of his hunting life introduced a few hundred head of Wapiti to Austria. . . . Now imagine this queerly mixed strip of river landscape as being bordered by vine-covered hills, brothers to those flanking the Rhine, from whose crests the two early medieval castles of Griefenstein and Kreuzenstien look down with serious mein over the vast expanse of wild forest and water. Then you have before you . . . the landscape which I consider the most beautiful on earth, as every man should consider his own home country."

Konrad received as a childhood present an aquarium and some fish, probably goldfish; he made himself a crude fishnet out of bent wire and a stocking, and "with such an instrument, caught, at the age of nine, my first Daphnia for my fishes, thereby discovering the wonderworld of the freshwater pond, which immediately drew me under its spell. In the train of the fishing net came the magnifying glass; after this again a modest little microscope, and therewith my fate was sealed; for he who has once seen the intimate beauty of nature cannot tear himself away from it again. He must become either a poet or a naturalist and, if his eyes are good and his powers of observation sharp enough, he may well become both."

His family seems to have accepted without demur his hobby of collecting stray animals, but there was never any question, seemingly, of the ultimate goal to which this biological curiosity should be directed. Like his older brother Albert who became a physician, and like their father, also a professor of medicine,

Konrad was aimed at medical school. In 1922, however, he went to America for a year's study at Columbia University, returning to Vienna to enroll in medical school. But during this period of medical training, an accidental event changed the course of Lorenz's life. He recalls the incident in his book *King Solomon's Ring*: "And so, as frequently happens with the great loves of our lives, I was not conscious of it at the time when I became acquainted with my first jackdaw. It sat in Rosalia Bongar's pet shop, which still holds for me all the magic quality of early childhood memories. It sat in a rather dark cage and I bought it for exactly four shillings, not because I intended to use it for scientific observations, but because I suddenly felt a longing to cram that great yellow-framed red throat with good food. I wished to let it fly free as soon as it became independent and this I recall I did, but with the true unexpected consequence that even today after the terrible war, when all my other birds and animals are gone, the jackdaws are still nesting under our rooftops. No bird or animal rewarded me so handsomely for an act of pity."

"It was entirely due to Jock [who turned out despite "his" name, to be a female] that in 1927 I reared fourteen young jackdaws in Altenberg. Many of her remarkable instinctive actions and reactions toward human beings, as substitute objects for fellow members of her species, not only seemed to fall short of their biological goal, but remained incomprehensible to me, and therefore aroused my curiosity. This awakened in me the desire to raise a whole colony of free flying tame jackdaws, and then study the social and family behavior of these remarkable birds."

This of course, was precisely what Lorenz proceeded to do. From the beginning, his work was strikingly original. In part, this was no doubt a function of the adventurous quality of Lorenz's mind, but he was also compelled to be original, forced to look upon his animals and their behavior with a new and inno-cent eye since at this time he had not yet received any orthodox zoological training. He had not been exposed to any of the traditional questions and answers that comprised the dialogue between Man and Nature. He was in medical school, where he had learned how to observe nature critically and attentively, but he was removed from the mind-warping armature of pro- fessorial guidance. He was impelled to frame his own, necessarily new, questions to nature; and the answers he received were also new, canted in a slightly different direction from the traditional responses that had become part of accepted dogma. Shortly after Jock acquired primary flight feathers, Lorenz writes that the bird "developed a really childlike affection for my person. It refused to remain by itself and called in desperation if ever I was forced to leave it alone. . . .

"Such a fully fledged young jackdaw attached to its keeper by all its youthful affections, is one of the most wonderful objects for observation that you can imagine. You can go outside with the bird and, from the nearest viewpoint, watch its flight, its method of feeding, in short all its habits, in perfectly natural surroundings, unhampered by the bars of a cage. I do not think that I have ever learned so much about the essence of animal nature from any of my beasts or birds as from Jock in that summer of 1925."

In 1927 Lorenz embarked on a more ambitious project. He writes that without his chance encounter with Jock in the pet store, without having been encouraged by his success in raising this single bird, he would never have dreamed of attempting to raise an entire colony by similar means. But, he writes, "as it was out of the question that I should act as a substitute for their parents and train each of these young Jackdaws as I had done Jock the previous year, and as through Jock I had become familiar with their poor sense of orientation, I had to think out some other method of confining the young birds to the place. After much careful consideration, I arrived at a solution which subsequently proved entirely satisfactory. In front of the little window of the loft where Jock now dwelt for some time, I built a long and nar- row aviary, consisting of two compartments which rested upon a stone-built gutter a yard in width, and which stretched almost the entire breadth of the house."

In the front compartment of this aviary he kept those birds that were scheduled to fly on any given day.   In the rear compartment he confined those nest mates of the free-flying birds that were not allowed to escape that day, and whose cries he hoped would serve as a beacon, directing the free flyers back to the aviary in time for the evening roosting.   In the beginning each of the birds was distinctively marked with a specially colored leg band, but in a short time these artificial distinctions became unnecessary.   "I know," he writes, "the characteristic facial expression of every one of those birds by sight.   I did not need to look first at their colored leg-rings.   This is no unusual accomplishment, every shepherd knows his sheep, and my daughter Agnes—at the age of five—knew each one of our many wild geese by their faces. Without having known all the jackdaws personally, it would have been quite impossible for me to learn the inner secrets of their social life."

This social life was built upon an order of rank.   "Every jackdaw of my colony," Lorenz writes, "knew each one of the others by sight. . . .   After some few disputes which need not necessarily lead to blows, each bird knows which of the others she has to fear and which must show respect to her.   Not only physical strength, but also personal courage, energy, and even the self-assurance of every individual bird are decisive in the maintenance of the pecking order.   This order of rank is extremely conservative.   An animal proved inferior, if only morally, in a dispute, will not venture lightly to cross the path of its conqueror, provided the two animals remain in close contact with each other. This also holds good for even the highest and most intelligent mammals.   A large *Nemestrinus* monkey bursting with energy, owned by my friend the late Count Thun-Hohenstein, possessed, even when adult, a deeply rooted respect for an ancient Javanese monkey half his size who had tyrannized him in the days of his youth."

One of the first and most important differences that Lorenz noted between his free-ranging birds and the domesticated or captive birds observed by Schjelderup-Ebbe was that the savagery of aggressive interactions was much modified. Under conditions of captivity, Lorenz writes that "this [savagery] is often carried so far that the wretched victim, bullied from all sides, is never able to rest, is always short of food, and if the owner does not interfere may finally waste away altogether." With his wild birds Lorenz discovered that an entirely different situation prevailed. "Those of the higher orders . . . are not aggressive towards birds that stand far beneath them; it is only in their relations towards their immediate inferiors that they are constantly irritable." This seems to be generally true about social rank interactions in most vertebrate societies, when the animals are permitted to live in freedom. It also seems to be true of human social rank interactions.

The actual happenings, which reveal by the conduct of the participants their relative social ranks, are often very difficult for the casual observer to analyze. We have each of us, doubtless, seen multitudes of these kinds of happenings among pigeons feeding in the park, seagulls at the seashore, and songbirds around garden feeding stations without ever realizing what was being enacted. Lorenz describes a typical instance: "A jackdaw sits feeding at the communal dish, a second bird approaches ponderously in an attitude of self-display with the head proudly erected, whereupon the first visitor moves slightly to one side, but otherwise does not allow himself to be disturbed. Now comes a third bird, in a much more modest attitude which, surprisingly enough, puts the first bird to flight; the second, on the other hand assumes a threatening pose, with his back feathers ruffled, attacks the latest comer and drives him from the spot. The explanation: The latest comer stood in order of rank midway between the two others, high enough above the first to frighten him and just so far beneath the second as to be capable of arousing his anger. Very high caste jackdaws are most condescending to those of the lowest degree and consider them merely as the dust beneath their feet; the self-display actions of the former are here

a pure formality and only in the event of too close approxima-
tion does the dominant bird adopt a threatening attitude, but he
very rarely attacks."

Lorenz also noted that high-ranking birds tended to intervene
in disputes between lower ranking birds. When "mediating
[such disputes], the arbitrator [invariably the highest ranking bird
involved in the dispute] is always more aggressive toward the
higher ranking of the two original combatants. . . . Since the
major quarrels are mostly concerned with nesting sites (in nearly
all other cases the weaker bird withdraws without a struggle),
this propensity of the strong male jackdaw ensures an active
protection of the nests of the lower members of the colony."

It is precisely the kind of social aggressions displayed in these
quarrels, which establishes the structure of vertebrate social
societies. The means by which structures come into being
have lately captured the interest of Lorenz and other ethologists.

They are not particularly interested in the kind of aggression
involved in prey-predator interactions. Usually the predator
strikes down his prey with the same dispassionate and economic
dispatch that a man might employ to pull down an apple from
the tree, or split open a melon with a knife. Though the gesture
of hacking open a melon is probably the same one that a cavalryman might use to hack open another man's head, the motive is
entirely different, and the emotional charge underlying the act
is also entirely different. A whole series of biochemical events
must occur in the brain of the cavalryman before the latter act can
be accomplished successfully.

These biochemical events are just beginning to be understood
at the present time; theories and hypotheses abound, but no firm
theoretical structure has as yet been widely accepted. We do
know that prime among the chemicals involved in acts of
intraspecific aggression among vertebrates are a set of amine
molecules manufactured profusely in the adrenal gland and somewhat less intensely by nerve endings throughout the body.
When certain of these chemicals are manufactured by nerve

endings, they suffuse the tissues immediately adjacent to the nerve and enable muscles to produce short bursts of exceptional energy. Also, in some as yet not very well understood way, they also affect mood, and are implicated in both the arousal and suppression of that state of mind we humans call rage.

In the past few years we have managed to gain some chemical control over these mood-altering molecules, and whole families of mood-altering drugs have appeared on the market, including the psychic energizers, tranquilizers, psychedelics, and so on. In humans, the neo-cortex, that newly evolved portion of the brain which serves as the site of reason and logic may stimulate the production of various catechol amines. This is not true of animals. No animal would stimulate itself into producing adrenalin over such abstract issues as religious, political, or scientific heresy, new art forms, or other such stimuli which we know, all too unfortunately, do serve to stimulate such production in humans.

Among animals these secretions are most usually stimulated by those massive changes within the endocrine system as a whole, which occur at regularly occurring cyclical periods and have to do with glandular anticipation of reproduction; they involve a great many specialized behaviors which may include flocking, migration, nest building, the appearance of special physical courtship displays (such as plumage changes among birds), the enactment of courtship rituals, *and* the manifestation of social aggression. Insofar as aggression is concerned, laboratory experiments have disclosed that, under certain special conditions, social rank orders have been altered by injecting subordinate members with male hormones, and watching them ascend the social rank order. Likewise, injecting a dominant with female hormones can, under certain circumstances, decrease his natural aggression and cause his "dethronement." But this is a very gross system of altering behavior, analogous to repairing a pocket watch with a ball-peen hammer. After this massive shock to its mechanism the watch may operate somewhat differently for a while, but the effects of this kind of treatment are neither necessarily permanent, nor particularly informative.

It is not yet known how much of this secretory activity is under neo-cortical control in humans.   If we can stimulate rage through the neo-cortex, it is at least conceivable that we can suppress it as well.   Lacking entirely a hominoid neo-cortex, lower animals have evolved some curious systems for the activation and supression of these chemical stimuli.   An investigation of these systems may disclose to us human beings how we may collectively control our aggressive instincts more effectively in time future than we have in the historical past.

Several years after he began studying jackdaws, Lorenz turned to the study of the wild greylag geese and mallard ducks that nested annually in the backwaters of the Danube near Altenberg. As was his custom, he preferred to hand rear a clutch of young birds which later became tame enough to allow close observation. But when he attempted this with geese, he noted a strange difficulty.   He writes that "if a greylag gosling is taken in human care immediately after hatching, all the behavior patterns which are slanted to the parents respond at once to the human being.   In fact, only very careful treatment can induce incubator-hatched greylag goslings to follow a goose mother.   They must not be allowed to see a human being from the moment they break their shell to the time they are placed under the mother goose.   If they do, they follow the human being at once."

Lorenz was intensely curious about this phenomenon.   It differed markedly from the behavior of young ducklings. Lorenz recalled that Otto Heinroth, one of the pioneers of early ethology had reported on this in 1910.   Lorenz quotes Heinroth's remarks: "I have often," wrote Heinroth, "had to try and place an incubator gosling with a pair that was leading very young birds.   In so doing, one meets all sorts of difficulties, which are typical for the whole psychological and instinctive behavior of our birds.   When you open the lid of an incubator where young ducklings have just broken their shells and dried off, they will at first duck and sit quite motionless. Then, when you try to pick them up, they scoot away with lightning speed. Quite often they jump to the floor and hide beneath various objects and one has a hard time getting hold of the tiny creatures."

This is precisely the kind of behavior that one would expect from young animals—a response to the threat implied by a large, strange, and most likely predaceous creature approaching them close on. But, as Heinroth reported, young goslings behave quite differently: ". . . they look at you without betraying any sign of fear; and if you handle them even briefly, you can hardly shake them off. They peep pitifully if you walk away, and soon follow you about religiously. I have known such a little creature to be content if it could just squat under the chair on which I sat a few hours after I had taken it from the incubator. If you then take such a gosling to a goose family with young of the same age, the situation usually develops as follows. Goose and gander look suspiciously at the approaching person, and both try to get themselves and their young into the water as quickly as they can. If you walk very rapidly, so that the young have no chance to escape, the parents at first regard the newcomer as their own, and show an inclination to defend it as soon as they see and hear it in human hands. But the worst is yet to come. It doesn't even occur to the young gosling to treat the two old birds as geese. It runs away, peeping loudly, and, if a human being happens to pass by, it follows him: it simply looks upon human beings as its parents." Heinroth writes that the young goslings look upon the first moving object they encounter upon entering the world "with the intention of stamping its picture accurately in their minds, for, as I have said before, these pretty fluffy creatures do not seem to recognize their parents instinctively as members of their own species."

Fascinated by this "Heinroth occurrence" Lorenz proceeded to study it in his usual sympathetic and systematic fashion. Eventually he constructed a theoretical explanation for it which he called "imprinting" and which has since become one of the

principle cornerstones of ethological theory. He realized that
there is an innate, genetically prompted desire on the part of
certain animals at certain crucial periods in their lives to define
elements in their environment as being potentially responsive to
their needs, no matter how inapplicable these definitions may
eventually prove to be.   In subsequent experiments, for example,
he imprinted on his goslings, a dog, a block of wood, a rubber
ball, each of which objects was defined by the goslings as being
mother.

This predisposition on the part of goslings to accept any moving
object as mother is exceptional and extreme.   It is very rare in
animals.   Far more commonplace among animals, and humans
as well, is a capacity for abstraction, or symbolization; an innate,
genetically prompted (in the case of animals) compulsion to
accept some unique and particular aspect of the thing desired as
being representative of the entire thing.   Such is the case with the
cattle tick, which defines a mammal solely in terms of the odor
of sweat.   Lorenz, investigating further the reluctance of ducks
to accept him as their foster mother, discovered that imprinting
may often (usually in fact) depend upon this kind of abstraction,
which triggers the imprinting process.   "Mallards," he writes,
"on the contrary, always refused to do this [follow him about as did
his goslings]. If I took from the incubator freshly hatched
mallards, they invariably ran away from me and pressed themselves
into the nearest dark corner.   Why?   I remembered that I had
once let a muscovy duck hatch a clutch of mallard eggs and that
the tiny mallards had also failed to accept this foster mother.   As
soon as they were dry, they had simply run away from her and
I had trouble enough to catch these crying, erring children.   On
the other hand, I once let a fat white farmyard duck hatch out
mallards and the little wild things ran just as happily after her
as if she had been their real mother.   The secret must have laid
in her call-note, for in external appearance the domestic duck was
quite as different from a mallard as was the muscovy; but what
she had in common with the mallard (which, of course, is the wild

progenitor of the farmyard duck) was her vocal expressions. . . .
The inference was clear, I must quack like a mother mallard in
order to make the little ducks run after me.   No sooner said than
done.   When, one Whit-Saturday, a brood of pure-bred young
mallards was due to hatch, I put the eggs in the incubator, took the
babies, as soon as they were dry, under my personal care, and
quacked for them the mothers' call-note in my best mallardese.
For hours on end I kept it up, for half the day.   The quacking
was successful.   The little ducks raised their gaze confidently
towards me, obviously had no fear of me this time and as, still
quacking, I drew slowly away from them, they also set themselves
obediently in motion and scuttled after me in a tightly huddled
group just as ducklings follow their mother.   My theory was
indisputably proved. . . .   Anything that emits the right quack
note will be considered as mother, whether it is a fat white pekin
duck, or a still fatter man."

Like the cattle tick, which will leap onto any object smelling of
sweat but will drop off if the object is not warm, the young
ducklings had another criterion for motherhood in addition to
quacking.   Lorenz writes of this: "In the beginning of these
experiments, I had sat myself down in the grass amongst the
ducklings, and in order to make them follow me, had dragged
myself, still sitting, away from them.   As soon, however, as I
stood up and tried, in a standing posture, to lead them on, they
gave up, peered searchingly on all sides, but not upwards towards
me, and it was not long before they began that penetrating piping
of abandoned ducklings that we are accustomed to call 'crying.'
They were unable to adapt themselves to the fact that their foster
mother had become so tall.   So I was forced to move along,
squatting low, if I wished them to follow me.   This was not very
comfortable . . . [yet] in the interests of science I submitted myself
literally for hours on end to this ordeal."

Lorenz's conceptualization of the principle of imprinting led to
the conception of the Innate Releaser Mechanism—or, as it has
come to be referred to in the jargon, the IRM.   The signal is

mechanical, as with insects; if the animal perceives the signal it "releases" a sequence of pre-planned behavior. Since Lorenz's original observations, IRM's have been discovered operative in numbers of bird and mammalian behavior sequences. For example, a stick pointed at a nestful of fledglings will release the gaping, chirping, food-soliticing behavior of the chicks.   Or, if a kite in the shape of a hawk's silhouette is flown head-first over a group of young birds, it will release fright and escape behaviors. That this is both innate and very highly specific has been proved in various controlled experiments.   The specificity of the releaser is incredible—for if the kite is hauled tail-first instead of head-first over the chicks, they will not respond with the appropriate behavior.

Such releaser mechanisms also operate among humans, though in our case the majority of these releasers are learned, culturally acquired, and not innate. One powerful example of such a releaser in our culture, is the sight of a woman's naked breast upon a man. It serves to stimulate, if not to release, sexual behavior.   But in other cultures, women can go about naked to the waist without eliciting any particular attention from men.   In the operation of these releaser mechanisms can be seen the precursors of symbols.   If a partial aspect of a literal object, such as the quacking sound of a duck mother, can be made to represent the totality of motherhood, then one can see how humans developed their capacity to allow part of an abstraction to represent the whole.   The symbol of the Christian cross can be seen as an evolutionary elaboration of such a releaser; this single visual object has come to represent an intricately complex constellation of abstractions, which includes doctrine, history, philosophy; an entire way of life.   Words can also be made into releasers; currently the word "communist" acts as just such a releaser for many Americans, releasing powerful emotions of fear and counter-aggression very much like the hawk-kite does for poultry.

The next important step in the development of ethological

doctrine came with the appreciation of the fact that if certain signals could serve to release behavior, they could also serve to suppress it.   However, an intermediate step intervenes here.   It was a Dutch zoologist by the name of Nicholas Tinbergen who was principally responsible for the articulation of this concept. The behavior chain begins with what Tinbergen calls "displacement activities."   Tinbergen describes them as follows: "It has struck many observers that animals may, under certain circumstances, perform movements which do not belong to the motor pattern of the instinct that is activated at the moment of observation.   For instance, fighting cocks may suddenly peck at the ground as if they were feeding.   Fighting European starlings may vigorously preen their feathers.   Courting birds of paradise wipe their bills now and then.   Herring gulls, while engaged in deadly combat, may all at once pluck nesting materials. . . ."

What happens during a displacement activity is that the animal is suddenly caught up in a dilemma, a conflict of purposes.   It is forcefully impelled to perform a particular action.   The body is charged with biochemical energy impelling muscular expenditure for release, while at the same time that this first compulsion calls for action, the animal is dissuaded from performing it by some inhibiting factor.   It then resorts to expending its energy in another, substitute way.   In animals these displacement activities are far more rigorously stereotyped and ritualized than they are in human beings, but vestiges of such displacement activities can be seen among us.   As it is with dogs and monkeys, human scratching is often a displacement activity, the motor response to puzzlement and conflict.   Most displacement activities derive from intention movements, and a clear-cut example of this can often be seen on urban street corners.   A man will partially raise his arm to hail a taxicab, but even before the gesture is fully completed, he realizes that it will be abortive, the hail will go unacknowledged.   Once the gesture has been begun, however, there is an innate compulsion to make use of it somehow.   The man is very reluctant to let his arm merely drop uselessly to his side. As a result, more often than not, he will make another gesture out of the original, he will smooth down his hair, or adjust his hat, or do something of this sort before letting his hand drop back to his side.

Tinbergen cites a more complex and stereotyped example of this kind of displacement activity from the repertory of the stickleback fish. "In many species, the male, even when strongly sexually motivated, is unable to perform coition as long as the female does not provide the sign stimuli necessary for the male's consummatory act. A male stickleback, for instance, cannot ejaculate sperm before the female has deposited her eggs in the nest. The appearance of a female strongly arouses his sexual impulses. When the female does not respond to his zigzag dance by following him to the nest—a common phenomenon in incompletely motivated females—the male invariably shows his nest ventilating movements, often of high intensity and long duration. The amount of displacement fanning can even be used as a very reliable measure of his sexual motivation. This doubtless is the reason why so many displacement activities are parts of courtship patterns. Male ducks, for example, regularly preen their plumage during courtship . . . herring gulls and other birds feed their mates during courtship; it is not at all improbable that this courtship feeding is displacement feeding."

Lorenz discusses briefly how such displacement activities can become incorporated into a stereotyped ritual. "There is a remarkable relationship between the animal's posture in the displacement situation and the direction of the displacement. The autochthonous excitation seems to spark over with particular frequency and ease into the nervous path of the behavior pattern whose initial phase the animal's posture fits at the time. Cocks threatening to attack, perform intention movements of leaping into the air against each other. In the course of this activity, they assume a posture where their bills are held close to the ground and then they perform displacement pecking." It is here, at this point, that Lorenz has taken his leap into an evolutionary development hypothesis that still remains comparatively controversial, although massive evidence is being accumulated to support his thesis. "In this," he writes, "and in a great many other cases, the form of a behavior pattern, which was primarily dependent upon two or more orientation reactions, each of them variable, has developed into a single, rigid, and in this new form, hereditary reaction. If I regard the development of a new hereditary coordination as the essence of ritualization, this is confirmed by the changes which both intention movements and displacement activities undergo in the course of evolution."

An example of the hereditary ritualization of such a displacement activity might perhaps be seen in the drumming display of the ruffed grouse. It is quite easy to visualize this drumming as having originated as the displacement of an intention movement to attack another male. Instead of beating his wings on his opponent, the grouse, afflicted with conflicting intentions, might well have beat his wings upon a log, and the selective advantages of this acoustical display may well have made the habit species-specific, embedded it into the genotype.

These kinds of hypotheses concerning the ways in which displacement activities may have evolved were a necessary theoretical precursor to any speculation about the evolution of aggression suppressors, for such suppressors have become ritualized into innate, stereotyped behaviors. In many cases, structual displays augmenting the behavior have evolved to make it even more distinctive and effective. When one observes these behaviors being enacted in vertebrates, it is often difficult to stretch the imagination to such a point as would permit any understanding of their beginnings, just as it is hard to see in the foreleg of a quadripedal reptile the process by which an eagle's wing emerged into being.

One can see some of these suppressors working more directly and simplistically in insect behavior. When William Morton Wheeler wanted to open his discussion of insect courtship rituals, and the commonplace institution of "nuptial gifts," he began by reporting the account of the mating of a praying mantis as recorded by Leland Howard in 1886. "A few days since," Howard began, "I brought a male of *mantis carolina* to a friend who had been keeping a solitary female as a pet. Placing them in the same jar, the male, in alarm, endeavored to escape. In a few minutes, the female succeeded in grasping him. She first bit off his left front tarsus, and consumed the tibia and femur. Next she gnawed out his left eye. At this, the male seemed to realize his proximity to one of the opposite sex and began to make vain endeavors to mate. The female next ate up his right front leg, and then entirely decapitated him, devouring his head and gnawing into his thorax. Not until she had eaten all of his thorax, except about three millimetres, did she stop to rest. All this while the male had continued his vain attempts to obtain entrance to the valvules, and now he succeeded as she voluntarily spread the parts open and union took place. She remained quiet for four hours and the remnant of the male gave occasional signs of life by a movement of one of his remaining tarsi for three hours. The next morning she had entirely rid herself of her spouse, and nothing but the wings remained. The female was apparently full fed when the male was placed with her, and had always been plentifully supplied with food."

Spider females often behave with similar predatory aggression toward their potential mates. Wheeler quotes the account of three British observers, Locket, Bristowe and Locket, who reported on the mating ritual of the spider *Pisaura mirabilis* as follows: ". . . a male was given a fly and placed in a box with a female. He proceeded to enwrap the fly with silk; and then walked about with it in a jerky fashion until presently the attention of the female was attracted, and she approached him. He held the fly out to her and, after testing it with her falces, she seized hold of it. The male then crept to a position almost beneath the female and inserted. . . . Throughout the mating, the female sucked the fly. . . .

"On the following day he was given a fly and enclosed in a

case where no female had ever been. He ate the fly without making any attempt to enwrap it."

The wrapping of this fly and the presentation of this "gift" to the female represents a simple kind of aggression suppressor. In this instance a substitute prey animal is provided by the male to the female so that he himself will not be taken for prey and eaten. Olfactory traces of the female's presence release this wrapping and presenting behavior in the male. This rather baroque behavior is ritualized; variations of it appear again and again throughout the insect world. In these instances the motivation of each partner to the transaction is eminently logical in human terms and the ritual easily comprehended.

Among vertebrates, however, the "logic" is more obscure, the suppressors work differently; they seem oddly one-sided and there is often no apparent benefit accruing to the aggressor animal to reward him *as an individual* for the unilateral suppression of his aggression. For the species, however, the suppression of this hostility is obviously and invariably adaptive to the survival of the species as a whole. Lorenz cites a clear example of this from the brood-tending behavior of the hen turkey.

As do many birds while brooding, the hen turkey becomes irritable and nervous at this time. Not being very intelligent, at least according to our anthropomorphic standards of intelligence, she is not innately capable of recognizing predators as such; she will attack any creature of a certain size which approaches her nest whether that creature be a rat, or a cat, or the stray chick from another species of bird. How does she recognize her own chicks? This would offhand seem simple enough—chicks are recognized as chicks because they look like chicks. However, Lorenz writes that this is not the case. The recognition process is at once more simple and paradoxically more complex than this.

Two associates of his, Margaret and Wolfgang Schleidt wished to test some turkey social interactions which they suspected were controlled by special call notes, and so to this end they caused several turkey hens to be deafened. They were horrified when they discovered these deaf hens attacking their own chicks as though they were predatory invaders of the nest area. Lorenz writes: ". . . a deaf hen, which has sat on false eggs for the normal incubating period, should be prepared to accept chicks, but if she is shown a day-old poult she does not react with maternal behavior: She utters no call notes, but if the baby approaches within yards, she raises her feathers defensively, hisses furiously, and as soon as the chick is within reach of her beak, she pecks as hard as she can. If we assume that the hen is in no way deranged other than in her hearing faculty, there can be only one interpretation of this behavior: she does not possess the slightest innate information as to what her chicks should look like, and she pecks at everything which moves near her nest, and which is not so big that her escape reaction transcends her aggression. Only the sound expression of the cheeping chick elicits innate maternal behavior and puts the aggression under inhibition."

Using normal turkeys to confirm the experiment from the other side, Lorenz installed small audio speakers into stuffed polecats (the wild turkey's commonplace nest predator) and dragged these stuffed dolls by means of thin wires toward brooding hen turkeys. When the speakers were emitting the taped cheeps of turkey chicks the hens ruffled their belly feathers, raised themselves off their nests and permitted the polecats to come into the nest as though they were chicks. When the speakers were silent, however, the hens attacked the polecats as they came within reach with deadly effectiveness.

The aggression-suppressive effect of immature plumage is easily understood in the context of another of Lorenz's citations —those events that customarily take place in a heron rookery.

Because of their large wingspan and long, ungainly legs, herons need some considerable area for landing and takeoff. Therefore nests in the rookery must be spaced widely enough so that neighboring birds will do nestling chicks no damage while landing, taking off, and caring for their own progeny. To this end, territorial quarrels within a heron rookery reach a considerable pitch of intensity. Mature adult herons remain continually hostile to their neighbors and if one heron comes within reach of his neighbor's outstretched neck and beak, he will be attacked. Herons are brought together in rookeries by their aggregating instinct, but they are segregated—healthily spaced—within the rookery by their fierce territoriality. Since immature birds will neither be mating nor nest-building, but merely entering the rookery to roost somewhere on a vacant branch, there is no adaptive advantage gained by subjecting them to the same kind of territorial attack as adult birds direct against one another. Their immature striped plumage serves as an aggression suppressor and allows them to settle inconspicuously among adult birds. Their nearby presence is, however, sufficiently disturbing that often the nearest adult will direct an attack upon another adult neighbor, who is out of range; this is testimony of the suppressive effect of the immature plumage. A direct attack against the immature intruder is suppressed or inhibited and redirected against another adult.

Love and hate, aggression and submission, need and satiety; one may see all these forces working themselves out quite simply and mechanically in the behavior of fish, reptiles, and birds. Ethological systems apply there; it is as one ascends the ladder of psychological complexity, as one observes animals up through the class of mammals, up through the order of primates, finally reaching man, that one finds what seems to be a progressive blurring of that which is innate, or given by the genetic heritage, and what is the individual response to individual experience.

But "blurring" is a poor metaphor; it seems more like an alternation, or oscillation—an oscillation between the historical past embedded within the very flesh and bone, and the perceived existential present. At each moment of our becoming (becoming older, wiser, other than what we were in the previous moment) we are being acted upon alternately by a pulse of autochthonous existence and a pulse of consciously perceived and intellectually evaluated existence; each alternating pulse modifying the next, so that, as with the sound of a flute, we are conscious, finally, only of the continuum, the thin, beautiful, and resonant sound of the self—the self, alive.

Linnaeus named us Homo sapiens, stressing, through that choice of appellation, all the arrogance and vanity of eighteenth-century scientific certainties.   In the two-hundred years ensuing between his time and ours, these certainties and vanities have become blunted and muted; we no longer see ourselves as belonging to one order of living being as separate from any other order of living being.   We are all interconnected by the fact of existence, the fact of life.   And this fact of life ties us irrevocably into the past as determining great parts of the present.

It is to the sound of this continuum, as well, that we must attend (so the new myths of biology advise us), for this is also the sound of the self, and we must listen in awe and humility, ever more aware of all that is contained within this sound, which we do not begin to understand.   The new biology tells us to listen to all the sounds of the self, attend it in all its movements, internal and external, and also attend those reflections of the self that surround us and enclose us within the web of life.   There is nothing new in this advice; every great myth of mankind has advised us to do precisely the same.   But somehow the metaphors of this new myth can make a claim of credibility on the contemporary imagination.

Being is transient, but life itself is immortal.

Notes

PART ONE
The Individual

CHAPTER 1

THE MOMENT OF BEING

3 It emerges from the egg not yet fully developed . . .
Jakob Johann von Uexküll, "A Stroll Through the Worlds of
Animals and Men," *Instinctive Behavior*, ed., Claire Schiller
(New York, International Universities Press, 1957).

7 "I remember discussions with Bohr . . ."
Werner Heisenberg, *Physics and Philosophy, the Revolution in
Modern Science* (New York, Harper and Brothers, 1958), p. 202.

8 ". . . the observation plays a decisive role in the event . . ."
Ibid., p. 202.

8 At one time we lived much more intimately with the moon . . .
Harold S. Burr, "Moon Madness," *Yale Journal of Biological
Medicine*, 16:249–256 (1944).
Harold S. Burr, "Electricity and Life: The Phases of the Moon
Correlated with Life Cycles," *Yale Science Magazine*, 18:5–6
(1944).
Harold S. Burr, and F. C. S. Northrup, "The Electro-Dynamic
Theory of Life," *Journal of Comparative Neurology*, 56:347–371
(1932).

12 ". . . the great factor in evolution is use and disuse . . ."
George Bernard Shaw, *Back to Methuselah* (London, Constable,
1921; Baltimore, Md., Penguin Books, Inc., 1939), p. 204.

13 ". . . respecting the perfectability of society . . ."
Encyclopaedia Britannica, 13th ed., article on Thomas Malthus
(Chicago, London, Toronto, Encyclopaedia Britannica, Inc.),
Vol. XIV, p. 717.

15 ". . . the time required for a given quantity of pure uranium to
decay into lead 206 is 4,560 years . . ."

Harold F. Blum, *Time's Arrow and Evolution* (Princeton, N.J., Princeton University Press, 1951).

16 He trained a trout to associate the color gray . . .
Jakob Johann von Uexküll, *op. cit.*

17 "All psychic process, feelings, and thoughts are invariably bound to a definite moment . . ."
Jakob Johann von Uexküll, *Theoretical Biology*, trans. by D. L. Mackinnon (London, Kegan, Paul and Trench, 1926), p. 15.

18 The brain of a cat is very fast, transmitting impulses . . .
John Paul Scott, *Animal Behavior* (Chicago, University of Chicago Press, 1958), p. 51.

18 "The hen is merely the egg's way of producing another egg."
Quoted by Colin S. Pittendrigh in *Behavior and Evolution: A Symposium*, Anne Roe and George Gaylord Simpson, eds. (Detroit, Wayne State University Press, 1958), p. 398.

20 There is, for example, one species of gecko lizard which flees . . .
Heini P. Hediger, *Studies in the Psychology and Behaviour of Captive Animals in Zoos and Circuses* (London, Butterworth's Scientific Publications, 1955; New York, Criterion Books, 1955), p. 42.

20 The characteristic flight of the tumbler pigeon . . .
Heini P. Hediger, *op. cit.*, p. 48.

21 This food goes into the rumen . . .
A. T. Simeons, *Man's Presumptuous Brain* (New York, E. P. Dutton & Co., Inc., 1961).

22 akinesis
Heini P. Hediger, *op. cit.*, p. 41.

23 "At first, they only discriminated against me when carrying a gun . . ."
Heini P. Hediger, *op. cit.*, p. 51.

27 "The crab uses its pincers for picking off small algae and cutting sponges . . ."
Adolph Portmann, *Animal Camouflage* (Ann Arbor, Mich., University of Michigan Press, 1959), p. 40.

28 ". . . old clothes, greasy cloths from the kitchen were the most sought after . . ."
Jane Goodall, "My Life Among Wild Chimpanzees," *National Geographic*, 124:272–308, Aug. 1963.

CHAPTER 2

CYCLICAL TIME

30 Many animal behaviors are dominated by the phase of the
   moon . . .
   C. Hauenschild, "Lunar Periodicity," *Cold Spring Harbor
   Symposium on Quantitative Biology*, 25:491–498 (1960).
34 Dr. Franz Halberg of the University of Minnesota . . .
   Franz Halberg, "Physiologic 24 hour periodicity in Human
   Beings and Mice; The Lighting Regimen and Daily Routines,"
   Photoperiodism and Related Phenomena in Plants and Animals, ed.
   Robert B. Withrow (Washington, D.C., American Associa-
   tion for the Advancement of Science, 1959).
35 Probably the first modern scientific investigations into circadian
   rhythms were conducted by a French astronomer . . .
   Quoted by Erwin Bünning in *Cold Spring Harbor Symposium on
   Quantitative Biology*, 25:1–10 (1960).
37 But one of the most fascinating and dramatic experiments on this
   subject was conducted by a zoologist, Dr. Frank A. Brown,
   Jr. . . .
   Frank A. Brown, Jr., "Response to Pervasive Geophysical
   Factors and the Biological Clock Problem," *Ibid.*, p. 57–72.
39 But the period itself, its rhythm and duration and "shape" of
   the cycle is jolted out of its sequence . . .
   Jürgen Aschoff, "Exogenous and Endogenous Components in
   Circadian Rhythms." *Ibid.*, p. 11–28.
40 Forel took his breakfast outside in his garden . . .
   August Forel, *Das Sinnesleben der Insekten* (Munich, Reinhardt,
   1910), p. 72.
40 Von Frisch's curiosity had been piqued by a series of papers
   published in 1910 . . .
   Karl von Frisch, *Bees, Their Vision, Chemical Sense, and Language*
   (Ithaca, N.Y., Cornell University Press, 1960), p. 4.
40 In order to test the color sense of honeybees, von Frisch began
   training wild bees to select small bowls of sugar water . . .
   Ibid., p. 6.
41 "When I wish to attract some bees for training experiments, I

NOTES

usually place upon a small table several sheets of paper which have been smeared with honey. . . ."
  Ibid., p. 53.
41 He devised a two-digit color code . . .
  Ibid., p. 54.
41 . . . "begins to perform what I have called a round dance. . . ."
  Ibid., p. 55.
41 . . . "run a short distance in a straight line while wagging the abdomen very rapidly from side to side . . ."
  Ibid., p. 70.
42 It only became apparent to von Frisch gradually . . .
  Ibid., p. 74.
43 "If," he writes, "the run points straight down, it means, 'Fly away from the sun to reach the food' . . ."
  Ibid., p. 78.
44 At any rate, he writes: "Light rays coming directly from the sun consist of vibrations that occur in all directions perpendicular to the line along which the sunlight travels. . . ."
  Ibid., p. 90.
44 In experiments which he conducted in 1948, he used a sheet of polaroid plastic "about six inches wide and twelve inches long, and every part of the sheet acted as an analyzer . . ."
  Ibid., p. 92.
45 Lindauer found that he could induce honeybees to dance at night . . .
  Martin Lindauer, "Time Compensated Sun Orientation in Bees", *Ibid.*, p. 369–375.
46 "Position and feeding time are memorized in the dark hive at any time. . . ."
  Ibid., p. 372.
46 "In 1955 I had the opportunity to carry out such an experiment [to prove] that the time-sense of bees is able to function independently of diurnal exogenous factors."
  Max Renner, "The Contribution of the Honeybee to the Study of Time-Essence and Astronomical Orientation," *Ibid.*, p. 361–368.
48 An Italian zoologist, Floriano Papi of the Zoological Institute of the University of Pisa, had become interested in the sand flea, *Talitrus saltador*.

Floriano Papi, "Orientation by Night: The Moon," *Ibid.*, p. 475-480.

50 "During the first and last quarters of the moon," Papi writes, "the angles [of the flea's course] assumed with the mirror are very different from those expected."
Ibid., p. 477.

51 "The first question which arises in sun-orientation is: what possible information given by the sun is really used by the compass animal?"
Wolfgang A. Braemer, "A Critical Review of Sun-Azimuth Hypotheses," *Ibid.*, p. 43.

54 As one of his associates later wrote of it, "this discovery of a directed migration in a restricted space allowed him to overcome the main difficulty that had hitherto hampered experimental work on long distance orientation, namely the long distance itself."
Klaus Hoffman, "Experimental Manipulation of the Orientational Clock in Birds," *Ibid.*, p. 379.

55 The Sauers began their work in 1958, capturing summer nestlings and raising them by hand until the autumn.
E. Franz G. Sauer, and Eleonore M. Sauer, "Star Navigation of Nocturnally Migrating Birds", *Ibid.*, p. 463-473.

57 Brown captured some mud snails and devised a contrivance to test their magnetic sense. . . .
Frank Brown Jr., Miriam F. Bennett, H. Marguerite Webb, "An Organismic Magnetic Compass Response," *Biological Bulletins*, 119:65-74 (1960).

57 Brown writes: ". . . the two orientations of the experimental magnetic fields gradually change their effectiveness as a clear function of the sun's angle."
Op. cit.

62 The pineal body is a small grey or white structure . . .
Julian I. Kitay, Mark D. Altschule, *The Pineal Gland: A Review of the Physiologic Literature* (Cambridge, Mass., Harvard University Press, 1954).
Wilbur B. Quay, "Histological Structure and Cytology of the Pineal Organ in Birds and Mammals," *Progress in Brain Research*, Vol. X (Amsterdam, 1965).
F. Tilney, L. F. Warren, "The Morphology and Evolutionary

Significance of the Pineal Body," *American Anatomical Memoirs*, 9:257 (1919).

62 ". . . sphincter which regulated the flow of thought."
Kitay, *op. cit.*, p. v.

62 . . . two monographs on the subject were published independently, one in German by H. W. de Graaf and one in English by E. Baldwin Spencer.
Tilney and Warren, *op. cit.*

63 In 1958 two zoologists, Robert Stebbins and Richard Eakin . . .
Robert Stebbins, Richard Eakin, "The Role of the Third Eye in Reptilian Behavior," *American Museum Novitates*, No. 1870 (1958).

64 The Latin writers named the pineal "glandula superior" . . .
Tilney and Warren, *op. cit.*

64 Strangely enough this tremendously important work was done by a dermatologist . . .
Aaron B. Lerner, James D. Case, Robert V. Heinzelman, "The Structure of Melatonin," *Journal of the American Chemical Society* 81:6084 (1959).

65 Elephants and whales have minute pineal glands. Walrus have huge ones . . .
Wilbur B. Quay, *op. cit.*

66 . . . in 1917 two American zoologists, Carey P. McCord and Francis P. Allen . . .
Carey P. McCord, Francis P. Allen, "Evidences Associating Pineal Gland Function with Alterations in Pigmentation," *Journal of Experimental Zoology*, 23:207–224 (1917).

68 "It's nice to be able to contribute to another man's field . . ."
A personal communication to author.

69 In 1898 there was published by a German physician, Otto Heubner . . .
Kitay, *op. cit.*, pp. vii, 174, 191.

69 In that year Virginia Fiske, working at Wellesley College . . .
Virginia M. Fiske, G. K. Bryant, Janet Putnam, "Effect of Light on the Weight of the Pineal Gland in the Rat," *Endocrinology*, 66:489–491 (1960).
Virginia M. Fiske, Judith Pound, Janet Putnam, "Effect of Light on the Weight of the Pineal Organ in Hypophysecto-

mized, Gonadectomized, Adrenalectomized or Thouracil-fed
Rats," *Endocrinology*, 71:130–133 (1962).

70 Julius Axelrod and Herbert Weissbach, working at the National
Institutes of Health . . .

Julius Axelrod, Herbert Weissbach, "Enzymatic O-Methylation
of N-acetylserotonin to Melatonin," *Science*, 131:1312–13
(1960).

70 It would seem that an Italian pharmacologist, V. Ersparmer of
the University of Rome . . .

V. Ersparmer, *Archivio di Scienze Biologiche*, 31(2):63–95 (1946).

70 Two years later Maurice Rapport, a hemotologist . . .

Maurice Rapport, A. A. Green, Irvine H. Page, "Crystalline
Serotonin," *Science*, 108:329 (1948).

72 It was a professor of Pharmacology at the University of Edinburgh who seems to have been the first to note . . .

John H. Gaddum, K. A. Hameed, "Drugs Which Antagonize
5-Hydroxytriptamine," *British Journal of Pharmacology*, 9:240
(1954).

Daniel X. Freedman, Nicholas J. Giarmin, "Brain Amines,
Electrical Activity and Behavior," *E.E.G. and Behavior*.

77n By means of a beautifully designed experiment, two German
physiologists, Eberhard Dodt and Ewald Heerd, demonstrated
that in frogs, at least, the pineal eye still serves as a wavelength
discriminator.

Eberhard Dodt and Ewald Heerd, "Mode of Action of Pineal
Nerve Fibres in Frogs", *Journal of Neuropsychiatry*, 25:(3) 405–
429 (1962).

CHAPTER 3

THE MOLECULAR MEMORY

86 His name is Fred Griffith and he was working in His Majesty's
Ministry of Health in London during the 1920's.

Fred Griffith (untitled article), *Journal of Hygiene*, 27:113 (1928).

90 It was only in 1944 that this was finally accomplished by three

physicians, Oswald T. Avery, Colin McLeod, and Maclyn McCarty, at the Rockefeller Institute.

> Oswald T. Avery, Colin McLeod, Maclyn McCarty (untitled article), *Journal of Experimental Medicine*, 79:137 (1944).

91 For example, Robert P. Levine, the distinguished Harvard geneticist, can now write flatly in an introductory genetics text: . . .

> Robert P. Levine, *Genetics* (Holt Rhinehart and Winston, 1962), p. 64.

94 For example, Moses Maimonides the Jewish philosopher . . .

> Moses Maimonides, *The Guide for the Perplexed*. Translated from the Arabic by M. Friedlander (Routledge and Kegan Paul, London, 1951), p. 161.

95 They describe the gene as "a hereditary determinant which, in its alternative forms, is responsible for differences in a particular trait."

> Ruth Sager and Frank J. Ryan, *Cell Heredity* (John Wiley, N.Y., 1961), p. 28.

98 After some study he discovered that his new, as he called them, "killer strain" secreted a substance into the medium that he called paramecin . . .

> Tracy M. Sonneborn, "Recent Advances in the Genetics of *Paramecium* and *Euplotes*," *Advances in Genetics*, 1:264–368 (1947).

101 "Traditional genetic procedures," Sager writes . . .

> Ruth Sager, "Genes Outside the Chromosomes," *Scientific American*, 212(2):70–79 (Jan. 1965).

104 McConnell himself describes what followed in their work: "It was . . ."

> James V. McConnell, "Memory Transfer via Cannabalism in Planarians," *Journal of Neuropsychiatry*, 3, S1, p. 48 (1962).
> *Cf.* Robert T. Thompson, James V. McConnell, *Journal of Comparative Physiological Psychology*, 48:1 (1955).

105 . . . in 1957 two biochemists, Roy John and William Corning . . .

> William C. Corning, E. Roy John, "Planarian Tail Learning Blocked by Ribonuclease," *Science*, 134:1363–1365 (1961).
> *Cf.* William Dingman, Michael Sporn, "Molecular Theories of Memory," *Science*, 144:27 (1964).

105 "John reasoned," writes McConnell . . .
   McConnell, *op. cit.*
106 "In 1957 when we got our first results on retention . . ."
   Ibid.
110 According to John Gaito . . .
   John Gaito, "DNA and RNA as Memory Molecules," *Psycho-logical Review*, 70:471–9 (1963).
110 They found "that administration of RNA to individuals with pre-senile, arteriosclerotic, and senile syndromes (with some degree of memory impairment) brought about memory improvement . . ."
   D. Ewen Cameron, Leslie Solyom, "Effects of Ribonucleic Acid on Memory", *Geriatrics*, 16:74–81 (1961).
   Also *cf.* Leonard Cook, Arnold B. Davidson, Dixon J. Davis, Harry Green, and Edwin J. Fellows, "Ribonucleic Acid: Effects on Conditioned Behavior in Rats", *Science*, 141:268–269 (1963).
   Also *cf.* Hjylmar Hydén, E. Egyházi, "Nuclear RNA Changes in Nerve Cells During a Learning Experiment in Rats", *Proceedings of the National Academy of Science*, 48:1366–1373 (1962).
110 "I believe," he writes, "that thinking, both conscious and unconscious, and short-term memory involve electro-magnetic phenomena in the brain interacting with the molecular patterns obtained from inheritance of experience."
   Linus Carl Pauling, *The Nature of the Chemical Bond,* and *The Structure of Molecules and Crystals: an Introduction to Modern Structural Chemisty* (Ithaca, N.Y., The Cornell University Press, 1960), p. 570.

PART TWO
The Population

CHAPTER 4
THE SPECIES PROBLEM

118 "If we were to place the various isolating mechanisms of animals according to their importance, we would have to place behavioral isolation far ahead of all the others. . . ."
Ernst Mayr, *Animal Species and Evolution* (Cambridge, Mass., Belknap Press, Harvard, 1963), p. 106.

118 "Ethological barriers are the most important isolating mechanisms in animals. . . ."
Ibid., p. 95.

118 "A shift into a new niche or adaptive zone is, almost without exception, initiated by a change in behavior. The other adaptations to the new niche, particularly the structural ones, are acquired secondarily. . . . Most recent shifts into new ecological niches are, at first, unaccompanied by structural modifications. Where a new habit develops, structural reinforcements follow sooner or later."
Ibid., p. 604.

120 As Loren Eiseley describes him, he "had a poetic hunger of the mind to experience personally every leaf, flower, and bird that could be encompassed in a single lifetime."
Loren Eiseley, *Darwin's Century* (New York, Doubleday & Company, Inc., 1958), p. 10.

123 Mayr writes: " . . . those four species are sufficiently similar visually, that they confuse not only the human observer, but also silent males of other species . . ."
Ernst Mayr, *op. cit.*, p. 17.

125 "San Francisco Bay, which keeps the prisoners of Alcatraz

isolated from the other inhabitants of California, is not an isolating mechanism . . ."
Ibid., p. 91.
125 A patient zoologist, Eugene R. Hall, after observing them for a long period, in 1951 published a 466-page paper which finally convinced his fellow taxonomists that the weasels of North America really belonged to only four separate species.
Eugene R. Hall, "American Weasels," *University of Kansas Publication of the Museum of Natural History* (Lawrence, Kans., University of Kansas Press, 1951), 4:1–466.
126 Malaria is "the greatest single destroyer of mankind . . ."
William Osler, *The Evolution of Modern Medicine* (New Haven, Conn., Yale University Press, 1923), p. 43.
127 The disease is considered partially responsible, also, for the decline of several great Mediterranean cultures.
Paul F. Russell, *Malaria* (Oxford, England, Blackwell's Scientific Publications, 1952).
127 Russell's estimate of total contemporary malaria cases . . .
Ibid., p. 7.
128 The general description of the plasmodium parasite which causes the disease and the description of its symptoms is taken from the following principal sources:
William Bispham, *Malaria: its Diagnosis, Treatment, and Prophylaxis* (Baltimore, The Williams & Wilkins Company, 1944).
Martin D. Young and G. Robert Coatney, "The Morphology, Life Cycle and Physiology of *Plasmodium malariae* (Grassi and Feletti 1890)" in *A Symposium on Human Malaria with Special Reference to North America and the Caribbean Region*, ed. by Forest Ray Moulton (Washington D.C. The American Association for the Advancement of Science Publication Number 14, 1941), pp. 25–29.
Reginald D. Manwell, "The Morphology, Life Cycle and Physiology of *Plasmodium vivax*", in *op. cit.*, p. 30–39.
S. F. Kitchen, "The Morphology, Life Cycle and Physiology of *Plasmodium falciparum*", in *op. cit.*, p. 41–46.
Aimee Wilcox and Lucille Logan, "The Detection and Differential Diagnosis of Malarial Parasites in the Schizogonous and Sporogonous Cycles", in *op. cit.*, p. 47–62.

129 Laveran's paper dated December 15, 1880, describes the following: "On the 20th of October last, while examining by microscope the blood of a patient suffering from malarial fever I observed in the midst of the red corpuscles the presence of elements which appeared to be of parasitic origin. . . ."
William Bispham, *op. cit.*, p. 5.

134 One lab specimen had been maintained in the lab for eighty-six days . . .
Russell, *op. cit.*

135 The general history of the anopheles controversy . . .
Nicholas H. Swellengrebel, Abraham de Buck, *Malaria in the Netherlands* (Amsterdam, Scheltema and Holkema Ltd., 1938).

136 For the general history of Marston Bates' expedition to Tirana . . .
Marston Bates, *The Natural History of Mosquitoes* (New York, The Macmillan Company, 1949).

136 Mosquito swarms being mistaken for fires . . .
Bates, *op. cit.*, p. 53.

137 Harrison Gray Dyar was a trained entomologist . . .
Bates, *op. cit.*, p. 54.

137 "The swarming habits of the common males at White Horse were constant and interesting. . . ."
Quoted by Bates, *op. cit.*, p. 54.

138 "A change from bright light to dim light, or from darkness to dim light was equally effective . . ."
Bates, *op. cit.*, p. 54.

139 Dr. Kennedy told Bates of having seen a "swarm of several hundred males . . ."
Ibid., p. 52.

139 "We once observed that a small swarm had formed over a piece of white paper in a cage . . ."
Ibid., p. 53.

140 "Swarms in the big cage . . . responded immediately to a low hum . . ."
Ibid., p. 56.

140 "In not a few respects," they write, "the sounds of mosquitoes we have tested are like bird calls."
Morton C. Kahn, William Celestin, William Offenhauser,

"The Recording of Sounds Produced by Certain Disease Carrying Mosquitoes," *Science*, 101:333 (1945).

142 Then "the male, having established contact with the female, the latter almost immediately settles. The male now hangs head downward, suspended solely by his terminalia . . ."
Bates, *op. cit.*, p. 61.

142 Most of the work in matching the egg to the adult mosquito which eventually appeared from within it was done by two zoologists, Lewis W. Hackett and Alberto Missiroli.
Bispham, *op. cit.*, p. 45.

143 Toward the end of the decade, another disturbing note was reported by two American entomologists, Robert Matheson and Herbert Sumner Hurlbut.
Robert Matheson, Herbert S. Hurlbut, "Notes on Anopheles Walkeri," *American Journal of Tropical Medicine*, 17:237–242 (1937).

CHAPTER 5

COMPATIBILITY

149 "The swarming usually begins with the appearance of a few males, readily distinguished by their red anterior segments and their white sexual segments darting rapidly through the water . . ."
Frank R. Lillie, Ernest Everett Just, "Breeding Habits of the *Heteronereis Form of Nereis Limbata* at Woods Hole, Mass.," *Biological Bulletins*, 24:147–169 (1913).

150 In 1951 a team of German biochemists attempted to analyze this (as they called it) "sex stuff" without too much success.
A. Hauenschild, C. Hauenschild, "Untersuchen über die stoffliche Koordination der Paarung des Polychäten *Grubea clavata*," *Zoologische Jahrbücher Abteilung für Allgemeine Zoologie und Physiologie der Tiere*, 62:429–440 (1951).

151 "He then entwines the female," Just writes, "and straightens out, thus clutching her in the twist of his body."
Edward Everett Just, "Breeding Habits of *Platynereis megalops*," *Biological Bulletins*, 27:201–213 (1914).

151 All the material concerning the introduction of the Gypsy Moth
    into Massachusetts, the statements of witnesses at the public
    hearings, the various tests performed by Fernald and his associates,
    the whole history of the plague is obtained from:
        Edward H. Forbush and Charles Henry Fernald, *The Gypsy
        Moth* (Massachusetts Department of Agriculture, 1896).
156 Two Government entomologists, C. W. Collins and Samuel F.
    Potts, were assigned the task of devising an extract of female
    scent to be used for bait.
        Charles W. Collins, Samuel Frederick Potts, "Attractants for
        the Flying Gypsy Moth as an Aid in Locating New Infestations,"
        U.S. Dept. of Agric. Technical Bulletin, #336 (1932).
156 Finally, in 1960 three chemists, Martin Jacobson, Morton Beroza,
    and William A. Jones, working at the Beltsville, Maryland, United
    States Government Agriculture Research Station succeeded.
        Martin Jacobson, Morton Beroza, William A. Jones, "Isolation,
        Identification, and Synthesis of the Sex Attractant of the Gypsy
        Moth," *Science,* 132:1011–12 (1960).
157 "It is certain," wrote Charles Darwin, "that there may be extra-
    ordinary mental activity with an extremely small absolute mass
    of nervous matter."
        Charles Darwin, *The Descent of Man* (London, John Murray,
        1871; New York, The Modern Library, Inc., 1936), p. 436.
160 Spieth writes, "the male moves to her rear . . ."
        Herman T. Spieth, "Mating Behavior and Sexual Isolation in
        Drosophila," *Behavior,* 3:105–145 (1951).
162 He therefore decided to concentrate his experiment "on the
    olfactory sense which in *Drosophila* is located in scent organs in
    the third segment of the antennae . . ."
        Ernst Mayr, "The Role of the Antennae in the Mating Behavior
        of Female *Drosophila,*" *Evolution,* 4: 149–154 (1950).
163 "The males display actively to both kinds of females," he writes,
    "perhaps even more to the operated flies . . ."
        Ibid.
163 "Spieth believes," Mayr writes, "that the fluttering wings serve
    like a reverse propeller, the wing an air stream toward the courted
    female . . ."
        Ibid.

164 He did not like the usage of the term organism because, as he wrote, "the organism is neither a thing nor a concept, but a continual flux or process . . ."

William Morton Wheeler, "The Ant Colony as an Organism," *Essays in Philosophical Biology* (Cambridge, Harvard Univ. Press, 1939), pp. 4 and 5.

165 "The attractant of a moth," they write, "is produced and secreted by certain glands just as is a hormone . . ."

Peter Karlson, Adolph Butenandt, "Pheromones (Ectohormones) in Insects," *Annual Review of Entomology*, 4:38–58 (1960).

166 A British apiarist, F. W. L. Sladen, is credited with discovering the function of Nasanov's gland in that year, 1902.

Robert Evans Snodgrass, *The Anatomy of a Honeybee* (Ithaca, N.Y., Cornell University Press, 1956), p. 116.

167 In 1919, shortly after he began working with honeybees, von Frisch noted that once his training boxes containing bowls of sugar water had been visited by bees from a hive, other bees were attracted to the same box.

Karl von Frisch, "Uber den Geruschsinn der Bienen," *Zoologische Jahrbücher Abteilung für Allegemeine Zoologie und Physiologie*, 3:37:1–238 (1919).

170 Ribbands trained marked bees from two separate but otherwise identical colonies to accept syrup from bowls within an open-topped box.

Ronald Ribbands, *The Behavior and Social Life of the Honeybee* (London, Research Association, Ltd., 1953), p. 174.

173 "When a colony of honeybees loses its Queen, the worker bees soon become aware of this fact (often within thirty minutes or less) and the behavior of the colony as a whole tends to change from a state of organized activity to one of disorganized restlessness."

Colin G. Butler, "The Method and Importance of the Recognition by a Colony of Honeybees (*A. mellifera*) of the Presence of its Queen," *Transactions of the Royal Entomological Society*, 105:11–28 (1954).

174 Butler's next step was to begin a detailed study of the various kinds of physical contact exchanged between the Queen and the members of her colony.

Colin G. Butler, "Extraction and Purification of Queen Bee Substance from Queen Bees," *Nature*, 184:1871 (1959).

175 The molecule, as he finally identified it, had a remarkable chemical resemblance to the ovary-inhibiting hormone of prawns . . .

Colin G. Butler, "Analysis of the Queen Bee Substance," *Transactions of the Royal Entomological Society*, 155:417–431 (1962).

CHAPTER 6

SPACE

180 "Darwin early pointed out," wrote von Uexküll, "that the earthworm drags pine needles into its narrow cave . . ."

Jakob Johann von Uexküll, "A Stroll Through the Worlds of Animals and Men" (see note for page 3 on page 275).

183 "It is proper," Willughby wrote in a letter to Ray in 1678, "for the nightingale at his first coming, to occupy or seize upon one place as its Freehold into which it will not admit any other nightingale but its mate."

Margaret Norse Nice, "The Role of Territory in Bird Life," *The American Midland Naturalist*, 26:441–487 (1941).

183 "It is impossible among a great many species of birds, for numerous pairs to nest close to one another, but individual pairs must settle at precisely fixed distances from each other . . ."

Ernst Mayr, "Bernard Altum and the Territory Theory," *Proceedings of the Linnaean Society of New York*, 45–6 (March, 1934).

185 It was in 1933 that two students of reptiles showed that territorial behaviors could be observed among lizards."

G. Kingsley Noble, H. T. Bradley, "The Mating Behavior of Lizards: Its Bearing on the Theory of Sexual Selection," *Annals of the New York Academy of Science*, 35:25–100 (1933).

185 In 1935 a German investigator, Fritz Bramstedt, noticed that after paramecia had been confined for an hour or so in a triangular container, when transferred by pipette into a circular container, they continued to swim in a triangular path."

Fritz Bramstedt, "Dressurversuche mit *Paramecium caudatum* und *Stylonychia mytilus.*" *Zeitschrift Vergleichende Physiologie,* 22:490–516 (1935).

186 In 1952 Beatrice Gelber also claimed to have demonstrated a sense of space in paramecia . . .
  Beatrice Gelber, "Investigations of the behavior of *paramecium Aurelia,*" *Journal of Genetic Psychology,* 88:31–36 (1956).

186 Subsequent work has indicated that paramecia may involuntarily exude various chemical "markers" which help them locate themselves in space.
  Earl Hanson, personal communication to the author.

186 This was finally accomplished by the distinguished American zoologist William Henry Burt who finally wrote a paper in the *Journal of Mammalogy* aptly entitled "Territoriality and Home Range Concepts as Applied to Mammals."
  William Henry Burt, "Territoriality and Home Range Concepts as Applied to Mammals," *Journal of Mammalogy,* 24: 346–352 (1943).

187 "Although other chipmunks often invaded her territory, she invariably drove them away . . ."
  Ibid.

187 "It seems highly probable that most mammalian females attempt to drive away intruders from the close vicinity of their nests containing young . . ."
  Ibid.

187 John T. Emlen, in 1948, during the course of a study of the Norway rat in the city of Baltimore . . .
  David E. Davis, John T. Emlen, Allan W. Stokes, "Studies of Home Range in the Brown Rat," *Journal of Mammalogy,* 29:207 (1948).

189 "One of their paths ran along the wall adjoining the wooden table opposite to that on which the next box was situated."
  Konrad Lorenz, *King Solomon's Ring* (New York, Thomas Y. Crowell Company, 1952), p. 109.

192 But once an animal has become accustomed to a territory, even the reduced territory of a zoo cage, it is hard indeed for it to adjust to a new territory.
  Heini P. Hediger, *Studies of the Psychology and Behaviour of*

Captive Animals in Zoos and Circuses (London, Butterworth's Scientific Publications, 1955; New York, Criterion Books, 1955), p. 42.

193 *et sequ.* Most of the general information cited in these next few pages concerning techniques employed by various mammals to demarcate territory have been derived from the following sources: Hediger, *op. cit.*

Heini P. Hediger, "The Evolution of Territorial Behavior," *The Social Life of Early Man*, ed., Sherwood H. Washburn (Chicago, Aldine Publishing Company, 1961).

Heini P. Hediger, *Saugetiere Territorien und Ihre Markierung* (Leiden, Netherlands, 1949).

Heini P. Hediger, "Die Bedeutung von Miktion und Defakation bei Wildtieren," *Schweizerische Zeitschrift für Psychologie*, 3:296–304 (1944).

198 Studies by various teams of baboon watchers, particularly the American anthropologists Sherwood Washburn, Irvin DeVore and the zoologist Stuart Altmann . . .

Sherwood H. Washburn, personal communication to author.
Irvin DeVore, personal communication to author.
Stuart A. Altmann, personal communication to author.
Stuart A. Altmann, "A Field Study of the Sociobiology of the Rhesus Monkeys, *Macacca Mulatta*," *Annals of the New York Academy of Science*, 102:338–435 (1962).

200 "The urge to mark and to hold territories against other males is very marked . . ."

L. T. Evans, "Behavior of Castrated Lizards," *Journal of Genetic Psychology*, 48:217–221 (1936).

201 John J. Christian of the Johns Hopkins School of Hygiene and David E. Davis of Pennsylvania State University, weighed the adrenal glands of mice after social rank had been established in a confined space.

John J. Christian, David E. Davis, "Relation of Adrenal Weight to Social Rank in Mice," *Proceedings of the Society of Experimental Biology & Medicine*, 94:728 (1957).

201 In 1958 Herbert L. Ratcliffe and Michael T. Cronin, knowing that arteriosclerosis is a common cause of death among zoo animals, reviewed the autopsy records of mammals and birds that died in

the Philadelphia Zoological Gardens over a period of forty years
from 1916 to 1956.

Herbert L. Ratcliffe, Michael T. Cronin, "Changing Frequency
of Arteriosclerosis in Mammals and Birds at the Philadelphia
Zoological Gardens," *Circulation*, 18: 41–52 (1958).

203 Christian writes: "During an intensive study of the periodic die-
off in Minnesota, [biologists] were able to demonstrate that a
very small number of deaths could be attributed to infectious
disease . . ."

John J. Christian, "The Adreno-Pituitary System and Population Cycles in Mammals," *Journal of Mammalogy*, 31:247 (1950).

204 Referring to Dr. Hans Selye's classic definition of shock . . .

Hans Selye, "The General Adaptation Syndrome and the
Diseases of Adaptation," *Journal of Clinical Endocrinology*,
6:117–230 (1946).

204 The investigators of the showshoe rabbit die-off "showed that the
convulsions in hares did not occur until the liver glycogen dropped
below 2 per cent demonstrating that they were caused by a progressive fall in the glucose reserves . . ."

Christian, *op. cit.*

204 In 1960 another investigator, John B. Calhoun . . .

John B. Calhoun, "A Behavioral Sink," *Roots of Behavior*, ed.,
Eugene Bliss (New York, Harper & Brothers, 1962).

PART THREE

Social Organization

CHAPTER 7

SOCIALITY

215 The following description of its behavior, while generally true
of all the eight or nine species of this class, refers particularly to
the species *Dictyostelium disocideum*, which was identified in
1935 by a soil chemist named Kenneth B. Raper . . .

John T. Bonner, "A Descriptive Study of the Development of
the Slime Mold Dictyostelium," *American Journal of Botany*,
31:175 (1944).

217 "When the anterior portion of a migrating plasmodium is
removed," he writes, "the decapitated body ceases migration, nor
does it respond to light."

Kenneth B. Raper, "The Communal Nature of the Fruiting
Process in the Acrasiaeae," *American Journal of Botany*, 27:437
(1940).

218 "In the apical tip of the pseudoplasmodium, a group of cells near
the tip becomes rounded off and enlarged . . ."

John T. Bonner, "Epigenetic Development in the Cellular
Slime Molds," *Symposia of the Society of Experimental Biology*,
Vol. 18 (Cell Differentiation) (New York, Cambridge Univer-
sity Press, 1963).

219 Bonner describes it somewhat differently: ". . . to accomplish
this transformation, the slug first points its tip upward and stands
on its end . . ."

John T. Bonner, "Differentiation in the Social Amoebae,"
Scientific American (December, 1959), p. 152.

219 He wrote in his description that "a founder varies somewhat
in appearance [from ordinary amoeba] at the time it became
active . . ."

Bryan M. Shaffer, "The Cells Founding Aggregation Centres in the Slime Mold *Polysphondylium Violaceum*," *Journal of Experimental Biology*, 38: 833–44 (1961).

220 The nature of the signal was a chemical—a gas.
John T. Bonner, "How Slime Molds Communicate," *Scientific American* (August, 1963), pp. 84–94.

221 "If this," he writes, "was immediately killed, and the culture at once returned to darkness, the residual cells did not re-aggregate..."
Bryan M. Shaffer, *op. cit.*

223 ... later (in 1922) he announced these findings, in German, to the world at large.
Thorlief Schjelderup-Ebbe, "Beitrage zur Sozialpsychologie des Haushuns," *Zeitschrift für Psychologie*, 88:225–52 (1922).

224 "Every bird is a personality ..."
Thorlief Schjelderup-Ebbe, "Social Behavior in Birds," *Handbook of Social Psychology*, ed., C. Murchison (Worcester, Mass., Clark University Press, 1935).

225 Friederich Nietzche is reported to have said that "over the whole of English Darwinism, there hovers something of the odor of humble people in need and in straits."
Robert Hofstadter, *Social Darwinism in American Thought* (Philadelphia, University of Pennsylvania Press, 1944), p. 25.

227 His British biographer, George Woodcock, writes, "... he was born into the highest rank of the Russian aristocracy ..."
George Woodcock, I. Avakumovic, *The Anarchist Prince* (London, Boardman & Company, 1949).

228 ... as he himself said, "... an immense influence which only grew through the years."
Roger Baldwin, ed., *Kropotkin's Revolutionary Pamphlets* (New York, The Vanguard Press, Inc., 1927).
Note: The biographical material on Kropotkin and all quotations are taken from either Woodcock and Avakumovic, or from Baldwin's edition of Kropotkin's pamphlets.

237 "Two aspects of animal life impressed me most during the journeys which I made in my youth in Eastern Siberia and Northern Manchuria ..."
Pëtr Kropotkin, *Mutual Aid: A Factor of Evolution* (New York, McClure Phillips, 1902).

239 He quotes Darwin as having written that under adverse conditions "those communities which included the greatest number of the most sympathetic members would flourish best and produce the greatest number of offspring."
Charles Darwin, *The Origin of Species*, 2nd ed. (London, J. Murray, 1860), p. 163.

243 He writes that long before his arrival at Woods Hole, *Arbacia* had "been much used in studies of various aspects of development..."
Warder Clyde Allee, *Cooperation Among Animals with Human Implications*, rev. ed. (New York, Henry Schuman, 1951).

245 "Colloidal silver," he writes, "that is, the finely divided and dispersed suspension of metallic silver, is highly toxic to living things, even the hardy goldfish..."
Ibid.

CHAPTER 8

AGGRESSION

248 Lorenz even goes so far as to say in his latest book *On Aggression* that "we do not know of a single animal which is capable of personal friendship and which lacks aggression."
Konrad Lorenz, *On Aggression* (New York, Harcourt, Brace & World, Inc., 1966), p. 148.

252 While ethology is not a new discipline, the name and need for it having been proposed by St. Hilaire in 1859 . . .
Warder Clyde Allee, A. E. Emerson, A. Park, T. Park, K. P. Schmidt, *Principles of Animal Ecology* (Philadelphia, W. B. Saunders Company, 1965), p. 42.

252 Konrad describes it in the following way: ". . . the virgin wilderness of this stretch of country is something rarely found in the very heart of Old Europe."
Konrad Zacharias Lorenz, *King Solomon's Ring* (New York, Thomas Y. Crowell Company, 1952), p. xvi.

253 Konrad received as a childhood present an aquarium and some fish, probably goldfish; he made himself a crude fishnet out of bent wire and a stocking, and "with such an instrument, caught,

at the age of nine, my first Daphnia for my fishes, thereby discovering the underwater world of the fresh water pond, which
immediately drew me under its spell. . . ."
Ibid., p. 3.

256 "Every jackdaw of my colony," Lorenz writes," knew each one
of the others by sight. . . . After some few disputes which need not
necessarily lead to blows, each bird knows which of the others
she has to fear and which must show respect to her . . ."
Lorenz, *On Aggression*, pp. 148-9.

258 When "mediating [such disputes], the arbitrator . . . is always more
aggressive toward the higher ranking of the two original combatants . . ."
Ibid., p. 150.

260 He writes that "if a greylag gosling is taken in human care
immediately after hatching, all the behavior patterns which are
slanted toward the parents respond at once to the human being."
Konrad Zacharias Lorenz, "Companionship in Bird Life,"
Instinctive Behavior; Development of a Modern Concept, pp. 103-4.

262 "Mallards," he writes, "on the contrary, always refused to do
this. If I took from the incubator freshly hatched mallards, they
invariably ran away from me and pressed themselves into the
nearest dark corner."
Konrad Zacharias Lorenz, *King Solomon's Ring*, p. 42.

265 For instance, fighting cocks may suddenly peck at the ground as
if they were feeding. Fighting European starlings may vigorously
preen their feathers.
Nicholas Tinbergen, *The Study of Instinct* (Oxford, England,
The Clarendon Press, 1951), p. 113.

266 "In many species the male, even when strongly sexually motivated,
is unable to perform coition as long as the female does not provide
the sign stimuli necessary for the male's consummatory act."
Ibid., p. 116.

268 "A few days since," Howard began, "I brought a male of the
Mantis carolina to a friend who had been keeping a solitary female
as a pet."
William Morton Wheeler, "The Kelep Ant and the Courtship
of its Mimic *Cardiacephala Myrmex*," *Foibles of Insects and Men*
(New York, Alfred A. Knopf, Inc., 1928), pp. 161, 164.

269 Two associates of his, Margaret and Wolfgang Schleidt, wished to test some turkey social interactions which they suspected were controlled by special call notes, and so, to this end, they caused several turkey hens to be deafened.

Konrad Zacharias Lorenz, *On Aggression*.

Index

Index